The Functional Fold

Suzi Jarvis
Anika Mostaert
Editors

The Functional Fold

Amyloid Structures in Nature

Pan Stanford Publishing

Published by

Pan Stanford Publishing Pte. Ltd.
Penthouse Level, Suntec Tower 3
8 Temasek Boulevard
Singapore 038988

Email: editorial@panstanford.com
Web: www.panstanford.com

British Library Cataloguing-in-Publication Data
A catalogue record for this book is available from the British Library.

The Functional Fold: Amyloid Structures in Nature

ISBN 978-981-4267-40-3 (Hardcover)
ISBN 978-981-4267-62-5 (eBook)

Contents

Chapter 1

The World of Functional Amyloid

Douglas M. Fowler[a] **and Jeffery W. Kelly**[b]

[a]*Department of Genome Sciences, University of Washington, Seattle, Washington, USA*

[b]*Department of Chemistry and The Skaggs Institute of Chemical Biology, The Scripps Research Institute, La Jolla, California, USA*

dfowler@u.washington.edu

Amyloid is a fibrous, β-sheet-rich protein aggregate that was discovered over a century ago in the context of Alzheimer's disease. In fact, a large number of degenerative diseases feature amyloid formation. Surprisingly, many recent studies have shown that amyloid also constitutes an important, emerging class of protein fibres that fulfil an array of functions in organisms ranging from bacteria to humans. This chapter will introduce functional amyloid and cover background topics including protein folding and aggregation, the structural features of amyloid, mechanisms of amyloid formation, and amyloid as a pathological entity. Thus, this chapter seeks to provide a background and perspective that will be helpful in reading the rest of this book.

The Functional Fold: Amyloid Structures in Nature

Edited by Suzi Jarvis and Anika Mostaert

ISBN 978-981-4267-40-3 (Hardcover), 978-981-4267-62-5 (eBook)

www.panstanford.com

1.1 Protein Misfolding and Aggregation

Proteins are linear biopolymers composed of amino acid monomers that carry out a wide range of functions. Proteins generally adopt well-defined, three-dimensional structures as a consequence of their sequence and their environment.[1] While numerous natively unfolded examples exist, proper protein function typically depends on forming the correct three-dimensional structure. This process is known as protein folding.

Protein folding is a highly complex conformational transformation that converts an ensemble of largely orderless protein structures (random coil) into an ensemble of compact, well-defined structures (globular fold). Protein structure is dictated by the specific amino acid sequence of a given polypeptide, the general features of the polypeptide backbone and the environment in which it folds including the aqueous solvent and the presence of chaperones, folding enzymes, and the like. Backbone conformational and hydrogen bonding preferences favour the formation of three types of generalized conformation-reverse turns, α-helices, and β-sheets. Thus, protein folding is a spontaneous, intramolecular process typically occurring on a short timescale that yields functional protein molecules.[2]

Despite eons of evolutionary fine-tuning, the protein folding process is always in competition with misfolding and aggregation. Like intramolecular protein folding, intermolecular aggregation is influenced by the conformational requirements of the amide backbone as well as the amino acid composition, the environment and the polypeptide conformation. Thus, aggregates with specific quaternary structures (α-helical or β-sheet) can be formed. Amyloid is a compact, fibrous, and β-sheet-rich protein quaternary structure that, until recently, was solely linked to a variety of degenerative diseases.

1.2 Structural Properties of Amyloid

Amyloid fibres are formed by the self-assembly of protein or peptide monomers into intermolecularly hydrogen bonded β-sheets (Fig. 1.1a,c).[3] The amyloid fold is sometimes called a cross-β-sheet quaternary structure because β-strands are oriented perpendicular to the fibre axis. The molecular structure of amyloid fibres has been extensively studied using numerous biophysical techniques for more than 40 years.

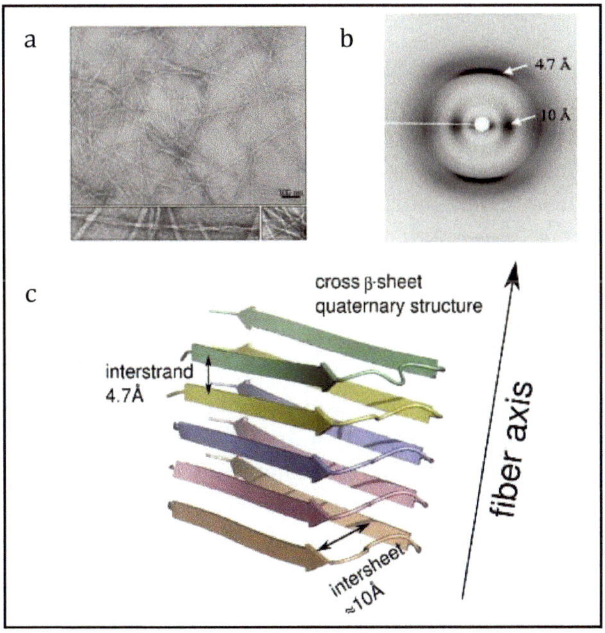

Figure 1.1 Amyloid fibres have a cross-β-sheet fold. Amyloid is a fibrous protein quaternary structure. (a) A transmission electron micrograph of negatively stained amyloid fibres formed from full length IAPP (reprinted with permission from Wiley-Liss, Inc., a subsidiary of John Wiley & Sons, Inc.; from Stromer *et al.*[4]). Higher magnifications (lower panels) reveal twisted rope and sheet like arrangements of individual protofilaments. (b) An X-ray fibre diffraction pattern from partially aligned Aβ (1–42) amyloid fibres associated with Alzheimer disease exhibiting the characteristic reflections at 4.7 and ≈10 Å (reprinted with permission of Wiley-Liss, Inc., a subsidiary of John Wiley & Sons, Inc., from Stromer *et al.*[4]). The meridional reflection at 4.7 Å results from the inter-strand repeats and the ≈10 Å equatorial reflection arises from inter-sheet packing. (c) Aβ (1–42) fibre structure obtained from solid-state NMR techniques and complementation mutagenesis methodology[5,6] reveals the characteristic cross-β-sheet amyloid structure (PDB ID 2BEG). Molecular graphics were produced with Pymol (DeLano, W.L. The PyMOL Molecular Graphics System (2002) on World Wide Web, http://www.pymol.org). See also Colour Insert.

Recent efforts have yielded both NMR-[5–7] and X-ray diffraction-based[8] high-resolution structures of peptide-derived amyloid fibres, revealing paired β-strands. These β-strands form in-register intermolecular interactions with identical β-strands via hydrogen bonding and side-chain interactions (Fig. 1.1c). While parallel, in-register cross-β assemblies seem to predominate based on the limited structural information available, both parallel and anti-parallel β-sheet orientations have been observed.[9]

Amyloid formation can be assessed using a variety of techniques *in vitro*. The fibrillar morphology of amyloid can be demonstrated using atomic force microscopy (AFM) or electron microscopy (EM). Far ultraviolet circular dichroism (CD) spectroscopy and Fourier-transform infrared spectroscopy (FTIR) have been used to demonstrate the β-sheet-rich quaternary structure of amyloid. Amyloid has distinctive reflections at 4.7 and 10 Å in its X-ray diffraction pattern because of its repetitive cross-β-sheet structure (Fig. 1.1b). The amyloidophilic, environment-dependent fluorophores Congo red and thioflavin S/T[10] can be used to reveal amyloid deposits in tissue using optical microscopy as well as to follow amyloid formation *in vitro* using fluorescence spectroscopy.

1.3 Amyloid Formation Mechanisms

Extensive biophysical studies of the mechanism of amyloid formation (amyloidogenesis) *in vitro* have revealed that amyloid is formed by the self-assembly of protein monomers that can occur by many different pathways. A nucleated polymerization mechanism has been invoked to explain the unusual lag in amyloid appearance observed for some amyloidogenesis reactions. In a nucleated polymerization, the lag phase corresponds to the slow, energetically unfavourable formation of the highest energy oligomeric species on the aggregation pathway, the nucleus (Fig. 1.2a). Fibre extension after nucleus formation is generally rapid and thermodynamically favourable. Another feature of a nucleated polymerization is that the monomer concentration must exceed a critical concentration below which amyloidogenesis will not occur. The nucleated polymerization mechanism is not the only mechanism of amyloidogenesis, as amyloid formation reactions that lack a lag phase have also been observed. These reactions are thought to proceed via a downhill polymerization in which all monomer addition events, including dimerizations, are energetically favourable (Fig. 1.2b). Many additional mechanistic paradigms will likely be discovered; one example is the recently described nucleated conformational conversion mechanism.[11]

Many, if not most, proteins can form amyloid under modestly denaturing conditions[12] suggesting that amyloid is a widely accessible, low-energy protein quaternary structure derived from partly denatured states. Amino acid composition can bias a given polypeptide towards amyloid formation. Hydrophobic sequences with a paucity of charge are particularly amyloidogenic, although the specific sequence determinants for amyloid formation remain poorly understood.[13]

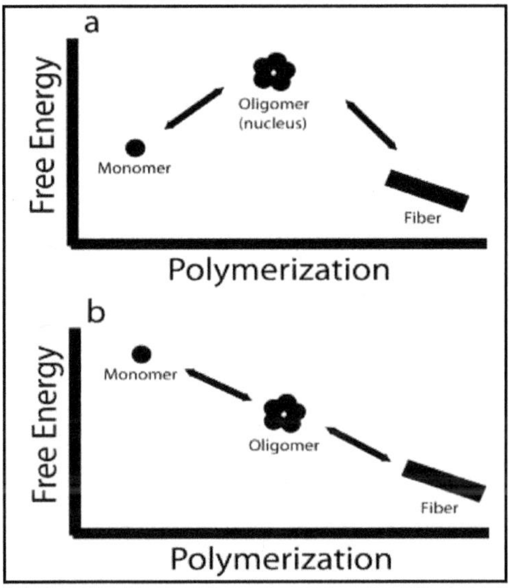

Figure 1.2 Amyloid formation can occur by many mechanisms, two of which are illustrated here. (a) In a nucleated polymerization, monomer assembly into a high-energy oligomeric nucleus is rate-limiting, with rapid polymerization to form fibres after nucleus formation. (b) In a downhill polymerization, all oligomerization steps are energetically favourable. The biophysical characteristics of a polypeptide dictate, in part, what mechanism will be observed. Conditions such as peptide concentration and solution composition can also modulate the mechanism. Our understanding of amyloidogenesis *in vivo* is poor; alternate mechanisms, influenced by chaperones and for example, are likely to be found.

Amyloidogenic proteins are diverse in sequence and share few characteristics—they are large and small, catalytic and structural, abundant and sparse. It should be noted that relatively little is known about mechanisms of amyloid formation *in vivo*. The mechanisms of amyloid formation *in vivo* are likely to be different from what is observed *in vitro* owing to the complex environmental conditions (molecular crowding, chaperones, aggregases, disaggregases, etc.) found inside living organisms.

1.4 Amyloid as a Pathological Entity

Amyloid was discovered a century ago in the context of Alzheimer's disease, which is characterized by neurodegeneration and extra-cellular amyloid formation by the Aβ peptide. Subsequently, amyloid has been associated with a wide variety of diseases including Parkinson's, Huntington's, the prion diseases, and type II diabetes (Table 1.1).[3]

Table 1.1 Amyloid diseases

Amyloid forming protein	Associated disease
Aβ precursor protein	Alzheimer's disease
Prion protein	Spongiform encephalopathies
Amylin	Type II diabetes
Apolipoproteins	Cardiac amyloidoses
Transthyretin	Senile systemic amyloidosis
Cystatin C	Familial dementia
Lactoferrrin	Corneal amyloidosis
Immunoglobulin	Myeloma-associated amyloidosis
Gelsolin	Familial amyloidosis
Insulin	Iatrogenic amyloidosis
β-Microglobulin	Hemodialysis-associated amyloidosis
Lysozyme	Familial amyloidosis
Lactadherin	Senile aortic amyloidosis
Fibrinogen α-chain	Familial amyloidosis

Amyloid diseases are characterized by the presence of fibrillar amyloid deposits composed primarily of one amyloidogenic protein. Genetic and pathological evidence suggest that amyloidogenesis causes amyloid diseases, although this so-called amyloid hypothesis remains unproven. Amyloid diseases affect a diverse array of organs and tissues because the location and extent of amyloid formation are defined by the unique physiological and biophysical parameters of the amyloidogenic protein involved.[14]

Interestingly, amyloid diseases can either arise spontaneously or be the result of mutation. In cases such as transthyretin amyloidosis, deposition of the wild-type protein can cause disease in an idiopathic fashion, especially in old age, while specific mutations lead to early-onset forms of the disease.[15] In cases such as gelsolin amyloidosis, disease appears to arise only from the presence of mutations in the gelsolin protein.[16]

A significant challenge for the amyloid field has been in the understanding of how the process of amyloidogenesis causes tissue damage. Amyloid, especially oligomeric amyloidogenic intermediates, have been shown to be toxic to cells in culture,[17,18] but the mechanism of this acute toxicity and its relevance to the slow, progressive tissue damage observed in amyloid disease remains elusive. A significant hindrance is the fact that amyloid encompasses a continuum of structures from small oligomers to fully formed fibres that are in equilibrium and therefore difficult to isolate and study independently. This has confounded efforts to positively identify a particular toxic amyloid structure and its mode of action. One theory suggests that pore-like amyloid oligomers capable of compromising membrane integrity might be responsible for the tissue damage observed in amyloid diseases.[19–21] Alternatively, Aβ oligomers appear to disrupt synaptic plasticity and have been proposed as an important causal factor of Alzheimer's disease pathology.[22]

1.5 Amyloid Is a Functional Protein Quaternary Structure

Given the toxicity of amyloid precursors and their apparent role in disease, as well as the ability of fibres to mechanically compromise organ function, the first reports of functional amyloid were quite surprising. Functional amyloid was identified initially in a number of lower organisms including bacteria,[1,23] fungi,[24–28] and insects[29–30] and more recently in algae[31] and humans[32] (Table 1.2). Additionally, several functional amyloid candidates have been reported,[33–35] although stronger evidence is needed to positively classify these as amyloid. Collectively, these discoveries led to the functional amyloid hypothesis, which states that organisms have evolved to take advantage of the predisposition of polypeptides to form amyloid, despite the fact that amyloid can be toxic. Thus, the amyloid fold is a structural motif that can be adapted to serve a variety of functions analogous to other folds including the β-sandwich, β-barrel folds.

Table 1.2 Known functional amyloid proteins

Kingdom	Species	Protein	Function	Supporting data
Bacteria	*Escherichia coli, Salmonella* spp.	Curli	Biofilm formation, host invasion	EM,[36] CR,[36,37] ThT,[36] CD[36,37]
	Streptomyces coelicolor	Chaplins	Modulation of water surface tension	CD,[23] ThT,[23] EM[23]
Fungi	*Podospora anserine*	Het-s	Regulation of heterokaryon formation	EM,[38,39] CD,[38,39] FTIR,[38,39]CR,[38] NMR[40]
	Saccharomyces cerevisiae	URE2p	Regulation of nitrogen catabolism	ED,[41] X-Ray,[41] CR,[24] EM,[24] ThT[42]
	S. cerevisiae	Rnq1p	Unknown	ThT,[43] EM[43]
	S. cerevisiae	Sup35p	Regulation of stop-codon read-through	EM,[25] CR,[25] CD,[25] X-ray[44]
	Most fungi	Hydrophobins	Fungal coat formation, modulation of adhesion and surface tension	AFM,[26,45] CD,[26,45] FTIR,[45] ThT, [28,45] NMR,[26,35] CR,[26,28] X-ray[35]
Animalia	Insects and fish	Chorion proteins	Structural and protective functions in the eggshell	CR,[46,47] X-ray,[29,46] EM,[29,46,47] FTIR,[46,47] CD[47]
	Nephila clavipes	Spidroins	Structural — spider silk	EM,[48] CD[48]
	Homo sapiens	Pmel17	Melanin biosynthesis	X-Ray,[32] CR,[32] EM,[32] ThT[32]

CD, ultraviolet circular dichroism spectroscopy; CR, Congo red binding; ED, electron diffraction; EM, electron microscopy; FTIR, Fourier-transform infrared spectroscopy; NMR, nuclear magnetic resonance spectrometry; ThT, thioflavin T binding; X-ray = X-Ray diffraction.

Given the relative simplicity of the amyloid fold, amyloid fibres are able to take on a stunningly wide range of functions. In bacteria, algae, and insect eggshells, amyloid fibres play a structural role; they enable colony formation, protect the organisms from the environment, and facilitate surface adhesion.[1,29,31] Amyloid has also been proposed to comprise the structure of spider silk, which has incredible mechanical strength.[30] Fungal amyloid fibres serve as epigenetic elements that function as drivers of phenotypic diversity and heterokaryon incompatibility.[49] In mammals, amyloid functions to facilitate the synthesis of melanin while mitigating the inherent chemotoxicity of melanin formation.[32] Furthermore, amyloid-like motifs could be important for regulating human hemostasis.[50]

To take advantage of amyloid for biological function, cellular and biochemical regulation is important to avoid toxicity.[51] Organisms appear to have evolved unique methods for regulating the amyloid formation process to avoid toxicity, although some common themes exist. For example, in *Escherichia coli* the formation of curli amyloid fibres appears to be nucleated by the CsgB protein outside the cell.[50] *Saccharomyces cerevisiae* regulate prion amyloid formation with an intracellular chaperone network.[49,50,52] Functional amyloid in humans appears to be regulated both by proteolysis to release the fibrillogenic fragment and by extremely rapid polymerization kinetics to prevent exposure of the cell to potentially toxic species.[50] In general, functional amyloid structures are found either outside the cell or contained within membrane-delimited vesicles.[50] This physical sequestration of amyloid presumably serves to protect the cell. All these regulatory schemes are likely necessary to avoid the toxicity of amyloid formation.

Amyloid seems likely to have appeared relatively early in evolutionary terms. Many polypeptide sequences can form amyloid under appropriate conditions, and functional amyloid is found in organisms ranging from bacteria to humans. Additionally, studies of the *Saccharomyces* spp. Sup35 prion protein suggest that the capacity for amyloid formation has been conserved for hundreds of millions of years.[53] These observations, combined with the fact that amyloidogenic proteins share little in the way of sequence or structural homology, suggests that amyloid is a primitive structure that likely existed for as long as living organisms have used polypeptides. Given its utility, simplicity, and self-seeding mechanism of formation, amyloid could have been a prominent fold early in the evolution of life.

1.6 Challenges and Future Directions

Although pathological amyloid was discovered over 100 years ago, the concept of functional amyloid is about 10 years old. In fact, only a handful of examples exist at this point. Since amyloid can fulfil a variety of functions and has, apparently, been conserved in evolution, it is likely that many more examples of functional amyloid exist. However, amyloid is a somewhat heterogenous mesoscale structure, making it difficult to detect and classify, particularly at the sub-cellular level. A central challenge for the field of functional amyloid is the development of an effective, high-throughput method for identifying more examples of functional amyloid. Imaging advances could offer such a solution, especially if amyloid-specific conformational antibodies prove robust. Identification of highly amyloidogenic polypeptide sequences using bioinformatics approaches could also guide *in vivo* investigations.

In addition to gaining insight into biology, the study of functional amyloid could lead to profound advances in medicine and nanomaterials (see chapter 10, this volume). Amyloid-related diseases, including Alzheimer's and Parkinson's, affect millions of people and are presently untreatable. A better understanding of how organisms regulate functional amyloid to prevent toxicity could point the way to treatments for these important neurodegenerative diseases. For example, recent evidence suggests that cells may contain generalized pathways for creating and removing amyloid-like aggregates.[54] Pharmacological control of endogenous systems for regulating the formation and degradation of amyloid could prove extremely useful in treating amyloid-associated diseases.[55]

Amyloid has remarkable mechanical and biochemical properties, exhibiting resistance to chemical denaturation and protease degradation as well as strength comparable to steel.[56,57] Many organisms have made use of these properties to effect a variety of biological processes; this suggests the possibility of using amyloid as a biomaterial. For example, spider silk, which likely contains amyloid-like motifs, has exceptional strength but is currently difficult to produce on industrial scales. A deeper understanding of how organisms manufacture amyloid could lead to the development of remarkably strong and light biomaterials.

In conclusion, amyloid appears to be an ancient, evolutionarily conserved structural motif or quaternary structure with an important role in many biological processes. Known examples of amyloid function range from protecting organisms from the environment and fulfilling structural roles to influencing chemical reactions within the cell. The discovery of functional amyloid challenges the long held belief that amyloid represents an inherently toxic structure associated only with aberrant misfolding and misassembly. Like any other protein structure, amyloid can be useful for function or can contribute to disease, particularly when its formation is unregulated. Further study and better methods of intracellular amyloid detection will likely reveal many more examples of functional amyloid. Understanding how amyloid structure, function, and regulation are related *in vivo* will have consequences for both treating amyloid-related diseases as well as for the use of amyloid as a biomaterial for related applications.

Acknowledgements

This work was supported by the National Institutes of Health (DK46335, NS50636 to J.W.K., and AG18917 to J.W.K. and William E. Balch), the Skaggs Institute of Chemical Biology and the Lita Annenberg Hazen Foundation. We acknowledge the contributions of our longtime collaborator William E. Balch for refining our ideas about functional amyloid. Evan Powers and Colleen Fearns provided assistance preparing this text.

References

1. Anfinsen, C. B. (1973) *Science,* **181**(4096), 223.
2. Daune, M. (1993) *Molecular Biophysics,* Oxford University Press, Oxford.
3. Chiti, F., and Dobson, C. M. (2006) *Annu. Rev. Biochem.,* **75**, 333.
4. Stromer, T., and Serpell, L. C. (2005) *Microsc. Res. Tech.,* **67**(3–4), 210.
5. Luhrs, T., Ritter, C., Adrian, M., Riek-Loher, D., Bohrmann, B., Dobeli, H., Schubert, D., and Riek, R. (2005) *Proc. Natl. Acad. Sci. U.S.A.,* **102**(48), 17342.
6. Petkova, A. T., Ishii, Y., Balbach, J. J., Antzutkin, O. N., Lesapman, R. D., Delaglio, F., and Tycko, R. (2002) *Proc. Natl. Acad. Sci. U.S.A.,* **99**(26), 16742.
7. Iwata, K., Fujiwara, T., Matsuki, Y., Akutsu, H., Takahashi, S., Naiki, H., and Goto, Y. (2006) *Proc. Natl. Acad. Sci. U.S.A.,* **103**(48), 18119.

8. Nelson, R., Sawaya, M. R., Balbirnie, M., Madsen, A. O., Riekel, C., Grothe, R., and Eisenberg, D. (2005) *Nature*, **435**(7043), 773.
9. Sawaya, M. R., Sambashivan, S., Nelson, R., Ivanova, M. I., Sievers, S. A., Apostol, M. I., Thompson, M. J., Balbirnie, M., Wiltius, J. J., McFarlane, H. T., Madsen, A. O., Riekel, C., and Eisenberg, D. (2007) *Nature*, **447**(7143), 453.
10. Westermark, G. T., Johnson, K. H., and Westermark, P. (1999) *Methods Enzymol.*, **309**, 3.
11. Serio, T. R., Cashikar, A. G., Kowal, A. S., Sawicki, G. J., Moslehi, J. J., Serpell, L., Arnsdorf, M. F., and Lindquist, S. L. (2000) *Science*, **289**(5483), 1317.
12. Guijarro, J. I., Sunde, M., Jones, J. A., Campbell, I. D., and Dodson, C. M. (1998) *Proc. Natl. Acad. Sci. U.S.A.*, **95**(8), 4224.
13. Bemporad, F., Calloni, G., Campioni, S., Plakoutsi, G., Taddei, N., and Chiti, F. (2006) *Acc. Chem. Res.*, **39**(9), 620.
14. Sekijima, Y., Wiseman, R. L., Matteson, J., Hammarstrom, P., Miller, S. R., Sawkar, A. R., Balch, W. E., and Kelly, J. W., (2005) *Cell*, **121**, 73.
15. Buxbaum, J. N., and Tagoe, C. E. (2000) *Annu. Rev. Med.*, **51**, 543.
16. Page, L. J., Huff, M. E., Kelly, J. W., and Balch, W. E. (2004) *Biochem. Biophys. Res. Commun.*, **322**(4), 1105.
17. Baglioni, S., Casamenti, F., Bucciantini, M., Luheshi, L. M., Taddei, N., Chiti, F., Dodson, C. M., and Stefani, M. (2006) *J. Neurosci.*, **26**(31), 8160.
18. Novitskaya, V., Bocharova, O. V., Bronstein, I., and Baskakov, I. V. (2006) *J. Biol. Chem.*, **281**(19), 13828.
19. Caughey, B., and Lansbury, P. T. (2003) *Annu. Rev. Neurosci.*, **26**, 267.
20. Lashuel, H. A., Hartley, D., Petre, B. M., Walz, T., and Lansbury, P. T., Jr. (2002) *Nature*, **418**(6895), 291.
21. Lashuel, H. A., Petre, B. M., Wall, J., Simon, M., Nowak, R. J., Walz, T., and Lansbury, P. T. (2002) *J. Mol. Biol.*, **322**(5), 1089.
22. Shankar, G. M., Li, S., Mehta, T. H., Garcia-Munoz, A., Shepardson, N. E., Smith, I., Brett, F. M., Farrell, M. A., Rowan, M. J., Lemere, C. A., Regan, C. M., Walsh, D. M., Sabatini, B. L., and Selkoe, D. J. (2008) *Nat. Med.*, **14**, 837.
23. Claessen, D., Rink, R., de Jong, W., Siebring, J., de Vreugd, P., Boersma, F. G. H., Dijkhuizen, L., and Wosten, H. A. B. (2003) *Genes Dev.*, **17**(14), 1714.

24. Thual, C., Komar, A. A., Bousset, L., Fernandez-Bellot, E., Cullin, C., and Melki, R. (1999) *J. Biol. Chem.*, **274**(19), 13666.

25. King, C. Y., Tittmann, P., Gross, H., Gebert, R., Aebi, M., and Wuthrich, K. (1997) *Proc. Natl. Acad. Sci. U.S.A.*, **94**, 6618.

26. Mackay, J. P., Matthews, J. M., Winefield, R. D., Mackay, L. G., Haverkamp, R. G., and Templeton, M. D. (2001) *Structure*, **9**, 83.

27. Coustou, V., Deleu, C., Saupe, S., and Begueret, J. (1997) *Proc. Natl. Acad. Sci. U.S.A.*, **94**(18), 9773.

28. Butko, P., Buford, J. P., Goodwin, J. S., Stroud, P. A., McCormick, C. L., and Cannon, G. C. (2001) *Biochem. Biophys. Res. Commun.*, **280**, 212.

29. Iconomidou, V. A., Vriend, G., and Hamodrakas, S. J. (2000) *FEBS Lett.*, **479**(3), 141.

30. Slotta, U., Hess, S., Spiess, K., Stromer, T., Serpell, L., and Scheibel, T. (2007) *Macromol. Biosci.*, **7**, 183.

31. Mostaert, A. S., and Jarvis, S. P. (2007) *Nanotechnology*, **18**.

32. Fowler, D. M., Koulov, A. V., Alory-Jost, C., Marks, M. S., Balch, W. E., and Kelly, J. W. (2006) *PLoS Biol.*, **4**, e6.

33. Diociaiuti, M., Polzi, L. Z., Valvo, L., Malchiodi-Albedi, F., Bombelli, C., and Gaudiano, M. C. (2006) *Biophys. J.*, **91**(6), 2275.

34. Si, K., Lindquist, S., and Kandel, E. R. (2003) *Cell*, **115**(7), 879.

35. Kamino, K., Inoue, K., Maruyama, T., Takamatsu, N., Harayama, S., and Shizuri, Y. (2000) *J. Biol. Chem.*, **275**(35), 27360.

36. Chapman, M. R., Robinson, L. S., Pinkner, J. S., Roth, R., Heuser, J., Hammar, M., Normark, S., and Hultgren, S. J. (2002) *Science*, **295**(5556), 851.

37. Cherny, I., Rockah, L., Levy-Nissenbaum, O., Gophna, U., Ron, E. Z., and Gazit, E. (2005) *J. Mol. Biol.*, **352**, 245.

38. Dos Reis, S., Coulary-Salin, B., Forge, V., Lascu, I., Begueret, J., and Saupe, S. J. (2002) *J. Biol. Chem.*, **277**, 5703.

39. Balguerie, A., Dos Reis, S., Ritter, C., Chaignepain, S., Coulary-Salin, B., Forge, V., Bathany, K., Lascu, I., Schmitter, J. M., Riek, R., and Saupe, S. J. (2003) *EMBO. J.*, **22**, 2071.

40. Ritter, C., Maddelein, M. L., Siemer, A. B., Luhrs, T., Emst, M., Meier, B., Saupe, S., and Riek, R. (2005) *Nature*, **435**, 844.

41. Baxa, U., Cheng, N., Winkler, D. C., Chiu, T. K., Davies, D. R., Sharma, D., Inouye, H., Kirschner, D. A., Wickner, R. B., and Steven, A. C. (2005) *J. Struct. Biol.*, **150**(2), 170.

42. Zhu, L., Zhang, X. J., Wang, L. Y., Zhou, J. M., and Perrett, S. (2003) *J. Mol. Biol.*, **328**, 235.
43. Sondheimer, N., and Lindquist, S. (2000) *Mol. Cell*, **5**, 163.
44. Kishimoto, A., Hasegawa, K., Suzuki, H., Taguchi, H., Namba, K., and Yoshida, M. (2004) *Biochem. Biophys. Res. Commun.*, **315**, 739.
45. de Vocht, M. L., Reviakine, I., Wosten, H. A., Brisson, A., Wessels, J. G. H., and Robillard, G. T. (2000) *J. Biol. Chem.*, **275**, 28428.
46. Kwan, A. H., Winefield, R. D., Sunde, M., Matthews, J. M., Haverkamp, R. G., Templeton, M. D., and Mackay, J. P. (2006) *Proc. Natl. Acad. Sci. U.S.A.*, **103**, 3621.
47. Iconomidou, V. A., Chryssikos, G. D., Gionis, V., Galanis, À. S., Cordopatic, P., Hoenger, A., and Hamodrakas, S. T. (2006) *J. Struct. Biol.*, **156**, 480.
48. Podrabsky, J. E., Carpenter, J. F., and Hand, S. C. (2001) *Am. J. Physiol. Regul. Integr. Comp. Physiol.*, **280**, R123.
49. Kenney, J. M., Knight, D., Wise, M. J., and Vollrath, F. (2002) *Eur. J. Biochem.*, **269**, 4159.
50. Uptain, S. M., and Lindquist, S. (2002) *Annu. Rev. Microbiol.*, **56**, 703.
51. Fowler, D. M., Koulov, A. V., Balch, W. E., and Kelly, J. W. (2007) *Trends Biochem. Sci.*, **32**, 217.
52. Hammer, N. D., Schmidt, J. C., and Chapman, M. R. (2007) *Proc. Natl. Acad. Sci. U.S.A.*, **104**, 12494.
53. Douglas, P. M., Treusch, S., Ren, H. Y., Halfmann, R., Duennwald, M. L., Lindquist, S., and Cyr, D. M. (2008) *Proc. Natl. Acad. Sci. U.S.A.*, **105**, 7206.
54. True, H. L., Berlin, I., and Lindquist, (2004) S. L. *Nature*, **431**(7005), 184.
55. Cohen, E., Bieschke, J., Perciavalle, R. M., Kelly, J. W., and Dillin, A. (2006) *Science*, **313**(5793), 1604.
56. Balch, W. E., Morimoto, R. I., Dillin, A., and Kelly, J. W. (2008) *Science*, **319**, 916.
57. Smith, J. F., Knowles, T. P., Dobson, C. M., Macphee, C. E., and Welland, M. E. (2006) *Proc. Natl. Acad. Sci. U.S.A.*, **103**, 15806.

Chapter 2

Hagfish Slime Threads

Douglas Fudge
Department of Integrative Biology,
University of Guelph Guelph, Ontario, Canada
dfudge@uoguelph.ca

Hagfish slime threads form the fibrous fraction of the defensive slime of hagfishes. The threads develop within specialized cells within the slime glands and consist of a dense bundle of the cytoskeletal elements known as intermediate filaments. The threads leave the hagfish's body in a condensed state, but quickly unravel into fine protein strands that are more than 10 cm long and approximately 2 μm in diameter. The proteins that make up the threads are dominated by α-helical secondary structure, but when acted upon by an external force, they stretch and adopt structures dominated by β-sheets. In this state, slime threads exhibit Congo red metachromasia and birefringence. These stretch-transformed threads differ from conventional amyloids in several ways. Most importantly, strands within β-sheets run parallel to the fibre axis, unlike the cross-β arrangement in typical amyloid fibrils. This arrangement means that stretched slime threads possess little in the way of "hidden length"

The Functional Fold: Amyloid Structures in Nature
Edited by Suzi Jarvis and Anika Mostaert

ISBN 978-981-4267-40-3 (Hardcover), 978-981-4267-62-5 (eBook)
www.panstanford.com

or "sacrificial bonds" and as such are quite stiff, without an obvious yield point. The study of hagfish slime thread mechanics has led to unexpected insights into the function of intermediate filaments in other systems, including the cytoskeleton of live cells and α-keratins. Although amyloid-like structures only appear in slime threads outside of the animal's body, it is possible that exceptional mechanical stresses could lead to amyloid-like structures within cells, with potential consequences for the development of conformational disease.

2.1 Introduction

Hagfishes are a group of ancient marine animals that pre-date the evolution of fishes. Phylogenetically speaking, they belong to the taxon Myxinoides, which is the sister taxon of Vertebrata. The 66 or so extant species of hagfishes[1] are found in an anti-tropical distribution and are typically found at depths greater than 100 m.[2] Hagfishes spend most of their time near or on the seafloor, with some species specializing in burrowing in soft sediments. Their lifestyle typically involves opportunistic scavenging as well as predation on benthic invertebrates like polychaete worms.[2] They have a low metabolic rate that allows them to go for months without feeding, but they can very quickly take advantage of carcasses that land on the bottom. In studies of scavenging ecology in deep sea habitats, hagfishes typically are one of the first species to arrive at baited traps.[3]

Hagfishes are remarkable creatures in many ways, including an ability to tie their body in an overhand knot to gain purchase on a particularly tough food source.[4] But hagfishes are best known for their superlative sliming abilities, which have earned them the common name "slime eel". Hagfishes have an astonishing ability to produce large volumes of slime when they are perturbed or stressed, and they do so far more quickly than the many marine species that appear to simply increase epidermal mucous secretions in response to stress. A typical mass of slime released from the Pacific hagfish *Eptatretus stoutii* has a volume of about a litre and can be formed in less than 100 ms[5,6] (Fig. 2.1).

Figure 2.1 Slime production by a Pacific hagfish in a bucket of sea water.[5] Photo by Christine Ortlepp. See also Colour Insert.

The slime originates in numerous (generally 150–200) slime glands that line both sides of the hagfish's body. The slime glands are approximately 2 mm in diameter and are packed with two different types of cells that make up the slime exudate. These are the gland mucus cells (GMCs) and the gland thread cells (GTCs).[7] GMCs are large secretory cells that are filled with countless vesicles containing condensed mucins. When a slime gland contracts, GMCs are ejected from the gland through the slime gland duct. During ejection, the GMC plasma membrane ruptures, releasing the mucin vesicles.

GTCs are large cells that manufacture a single protein thread that in mature cells takes over the vast majority of the volume of the cytoplasm. Like GMCs, GTCs undergo holocrine secretion from the gland, and also lose their plasma membrane during their short journey through the narrow slime gland duct. As it exits the slime gland, the slime exudate is a thick suspension of mucin vesicles and coiled thread cell polymers, also called "skeins" due to their resemblance to balls of yarn (Fig. 2.2).

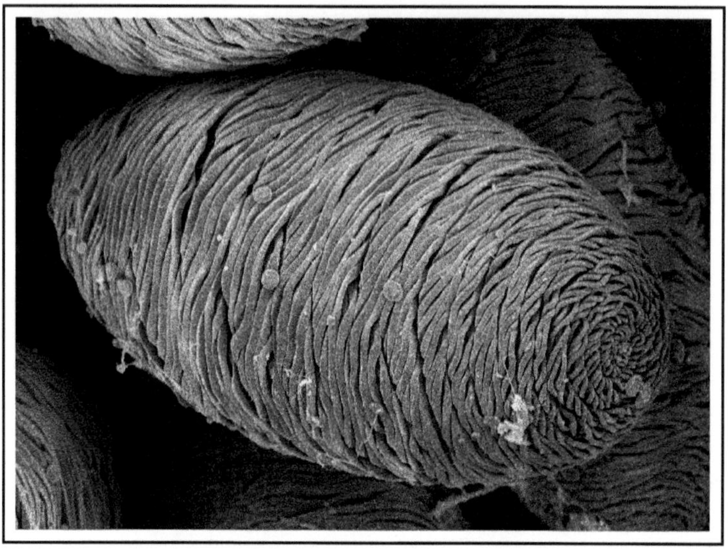

Figure 2.2 SEM of a mature gland thread cell (GTC). GTCs lose their plasma membrane as they are ejected through the slime gland duct and into the surrounding sea water. The vast majority of the cytoplasmic volume in these cells is occupied by a single protein polymer that consists of a dense bundle of 10 nm intermediate filaments.

Upon turbulent mixing with sea water, however, the slime quickly disperses and undergoes a remarkable transformation from a small volume of viscous exudate into a much larger volume of nearly transparent slime. Without turbulent mixing, slime exudate transferred into sea water (or even ejected from a live hagfish) will not form "proper" slime and will remain as a small coherent blob.[6] The exact mechanisms that are at play during the transformation of slime exudate into fully expanded slime are not yet fully understood; however, some clues are available.

One thing that occurs during this transformation is the complete unravelling and elongation of thread cell skeins into fibres that are more than 10 cm long.[5,8] This is an impressively long structure for something built within a single cell (although admittedly not as impressive as chromosomal DNA, which is 500 times narrower in width, almost 20 times longer, and packed into a structure that is only 1/20,000 the volume of a GTC). Another thing that occurs is the hydration and swelling of mucin vesicles as they are exposed to sea water.[9] Contributions of the thread and mucin vesicle components are synergistic—neither alone can produce anything resembling the native slime.[10] Further, the addition of the disulphide bond reducer dithiothreitol to sea water delays the formation of the slime and interferes with slime cohesiveness when slime exudate is added.[5,10] This effect is most likely brought about by a disruption of the disulphide bonds that hold mucin molecules into a coherent network, although the details of this mechanism are still currently unknown.

The mechanical properties of the slime are highly time- and deformation-dependent, and in this way hagfish slime may best be understood as an event rather than a conventional material. At first glance, an average mass of slime secreted by a Pacific hagfish appears to have a volume of about a litre, but this value depends greatly on the speed and manner in which the slime is collected. If the slime is quickly lifted out of water (Fig. 2.1), it brings with it a great deal of this water, but if the slime is held in air, the majority of water will eventually run free.[5] The dramatic egress of water from the slime suggests that the large volume of slime produced is mostly bulk sea water that is entrained in minute spaces within the slime. This conclusion is further supported by a simple model of slime morphology that demonstrates that it is not physically possible for the mucin and thread components to immbolise a litre of sea water.[5]

Figure 2.3 Experiments with isolated fish heads demonstrate that hagfish slime is effective at clogging gills.[6] Photo by Jeanette Lim. See also Colour Insert.

What kind of function could such an unusual material (or event) fulfil for the hagfish? One hypothesis is that the slime acts as a deterrent to gill breathing predators. Our data are consistent with this idea—the slime does not appear adapted to binding sea water, but rather to slowing it down to a trickle, which would have devastating effects on the function of fish gills.[6] Indeed, it has been shown more directly that hagfish slime is capable of dramatically increasing the resistance of gills in experiments with isolated fish heads[6] (Fig. 2.3).

2.2 Are Hagfish Slime Threads "Disposable Amyloid"?

One of the keys to the unique properties of hagfish slime is the rapid deployment of the slime threads from a tight, coiled state to one in which they are unravelled and extended. The presence of the threads imparts long-range order to the slime and allows it to remain as a semi-coherent mass even at the extremely low concentrations of protein and glycoprotein in mature slime (approximately 40 mg dry weight/l). In their native state, the slime threads are 1–3 μm in diameter and more than 10 cm long,[5,8] but they have the ability to stretch to more than three times this length when they are loaded in tension.[11] It is under these conditions that their constituent proteins take on an amyloid-like structure.

The amyloid-like nature of hagfish slime threads was first detected by staining the threads with Congo red, thus revealing amyloid-like metachromasia and birefringence.[5] Further investigations showed that only stretched slime threads exhibit these properties, while unstretched, native threads do not. These experiments led to further exploration of what kind of deformations of the slime are required to bring about the transformation of slime thread proteins into an amyloid-like state. The answer is that hagfish slime, as produced in a still aquarium, results in little Congo red metachromasia by slime threads, as long as the slime is handled gently when it is isolated and stained. If the slime is subjected to a mild mechanical stimulus, as might occur when a hagfish or other animal thrashes within it, most of the threads are transformed into a state in which they exhibit Congo red metachromasia (Fig. 2.4).[5]

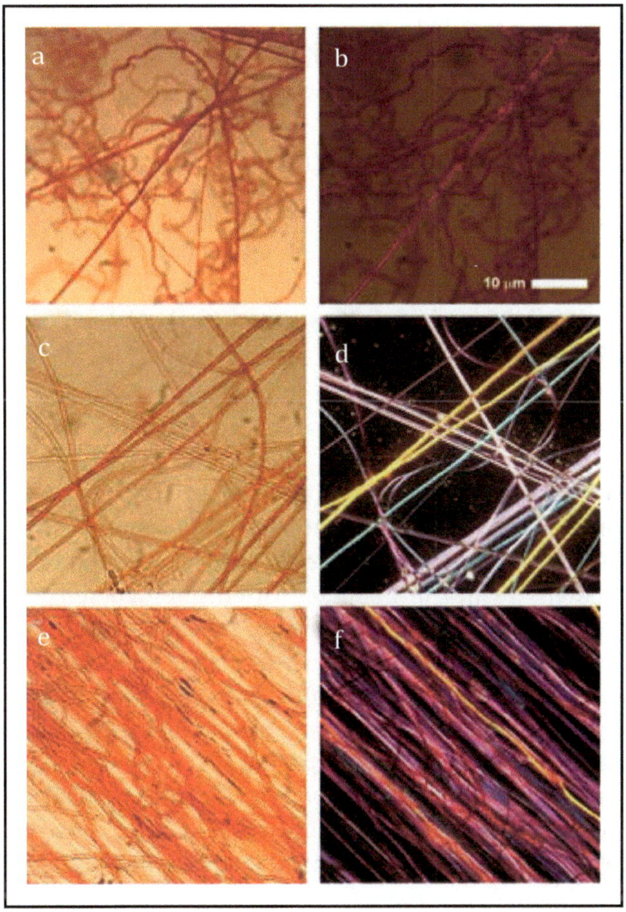

Figure 2.4 Congo red staining of whole hagfish slime produced in aquarium.[5] Images on the left are brightfield, and images on the right are taken with crossed polarizers. Images a and b are from slime that was handled gently, and images c–f are from slime that was purposefully deformed, as might occur when a hagfish or predator thrashes within it. Congo red metachromasia and birefringence is shown in d and f. See also Colour Insert.

These experiments demonstrate that the amyloid-like nature of hagfish slime threads only occurs outside of the hagfish's body and only when the slime is suitably deformed by an external force. This represents a case of functional amyloid that operates outside of the body of the organism that makes it.[12,13] This observation reinforces the emerging idea that amyloid-like protein structures may be useful for constructing biomaterials and adhesives with unique mechanical properties, several of which are highlighted in this book (see chapters 8 and 10, this volume). It also suggests that the advantages of employing structural proteins with amyloid motifs within an animal's body may be outweighed by the physiological risks of amyloidogenic disease.

2.3 Slime Threads Are Bundles of Intermediate Filaments

To understand the amyloid-like nature of stretched slime threads, it is worthwhile to take a step back and examine the kinds of proteins that make up the threads. Early work on the ultrastructure of slime threads, especially in immature GTCs within the slime gland, demonstrated that the bulk of the thread is composed of filaments approximately 10 nm in diameter. Often interspersed with these filaments are microtubules or microtubule doublets, although these become rarer as the thread cell matures and the 10 nm filaments become more plentiful. In the mid-1990s, Koch *et al.* established that the proteins that make up these 10 nm filaments belong to the "intermediate filament" family of proteins.[14]

Along with F-actin and microtubules, intermediate filaments are important components of the cytoskeleton of most metazoan cells. In living cells, cytoplasmic intermediate filaments often form a cage of filaments around the nucleus that radiate out toward the cell periphery, where they are anchored in desmosomes and hemidesmosomes.[15] Intermediate filaments are also noteworthy because they make up the fibrous phase of the fibre-reinforced composite materials known as α-keratins.[16] α-Keratins consist of intermediate filaments embedded in an elastic protein matrix, that make up not only the soft and pliable outer layers of epidermis in amniotes (the stratum corneum), and also a diverse array of harder epidermal appendages in mammals, such as hairs, nails, horns, claws, and the baleen plates of mysticete whales.[17]

Intermediate filaments comprise one of the most diverse gene families in humans, with at least 70 expressed genes now identified belonging to six different types.[18] Types I and II are the acidic and basic keratins proteins which are coexpressed as obligate co-polymers in epithelia and α-keratins. Type III intermediate filaments are expressed in cells of mesenchymal origin and include desmin (in muscle), vimentin (in embryonic cells and a variety of other cells such as fibroblasts and endothelial cells), and glial fibrillary acidic protein (in glial cells). Type IV intermediate filaments are the neurofilaments, and these are important components of the neuronal cytoskeleton, especially within axons and dendrites. Type V intermediate filaments are the nuclear lamins and these are found exclusively in the nucleus as a basket-like structure just under the nuclear envelope. Type VI intermediate filaments include nestin and paranemin as well as two proteins found in the vertebrate oculuar lens, filensin, and phakosin.

The proteins that make up the 10 nm filaments within hagfish slime threads clearly belong to the intermediate filament family, and have been cautiously classified as "keratin-like".[14] Slime thread intermediate filaments assemble from equal molar ratios of two proteins (named α and γ) and in this way they are heteropolymers like the keratins. There are some important differences to consider when comparing "true keratins" and the intermediate filaments in hagfish slime threads. Hagfish α and γ have a higher threonine content than other keratins (approximately 12% vs. 4%), and they have a linker segment (L12) that is 21 residues long, rather than the 16–17 residues found in keratins.[14] Slime threads are unique in that they represent the only known case of a secreted intermediate filament, and this may have released these proteins from certain kinds of selective pressures that have shaped the evolution of the other intermediate filament proteins. In spite of these differences, there is mounting evidence that the mechanical properties of slime thread intermediate filaments are similar to more conventional intermediate filaments.[19]

2.4 Mechanical Properties

One of the themes of this book is that amyloid is a mechanically useful motif for building protein materials. What then are the mechanical properties of slime threads, both before and after they are converted to amyloid? Overall, slime threads are remarkably extensible in water, being able to stretch on average to 220% strain before failing[11] (Fig. 2.5).

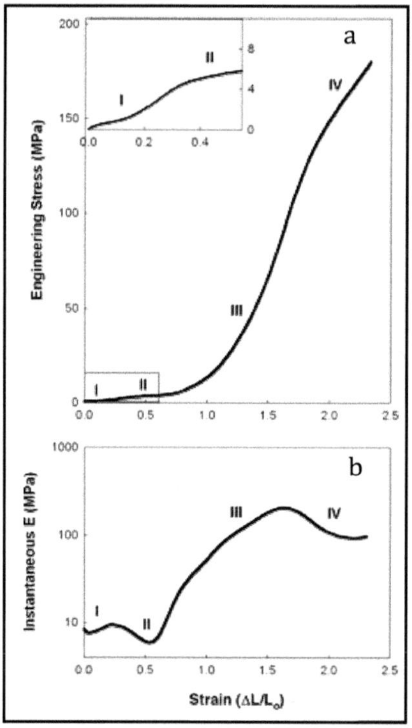

Figure 2.5 Tensile mechanics of slime threads. (a) Stress–strain curve for a single slime thread measured using a glass rod force transducer apparatus. (b) A plot of the instantaneous slope of the curve in (a), which makes is easier to see the different mechanical regimes denoted by the roman numerals. With permission from *Biophysical Journal.*[11]

This means that the threads can stretch to more than three time their original length before they break. For individual slime threads, this corresponds to a breaking length of 34 cm. The extensibility and stiffness of materials are typically inversely correlated, so it is not surprising that hydrated slime threads have a fairly low stiffness of approximately 6 MPa, which is similar to that of rubber-like biomaterials.[20,21] Comparable to rubber, slime threads are able to stretch reversibly to considerable strains. Unlike rubber, however, which behaves elastically at all strains up to its breaking point, slime threads yield at a strain of approximately 35%. Strains that exceed the yield strain are not reversible and lead to an increase in the resting length of the thread after it is unloaded[11] (Fig. 2.6).

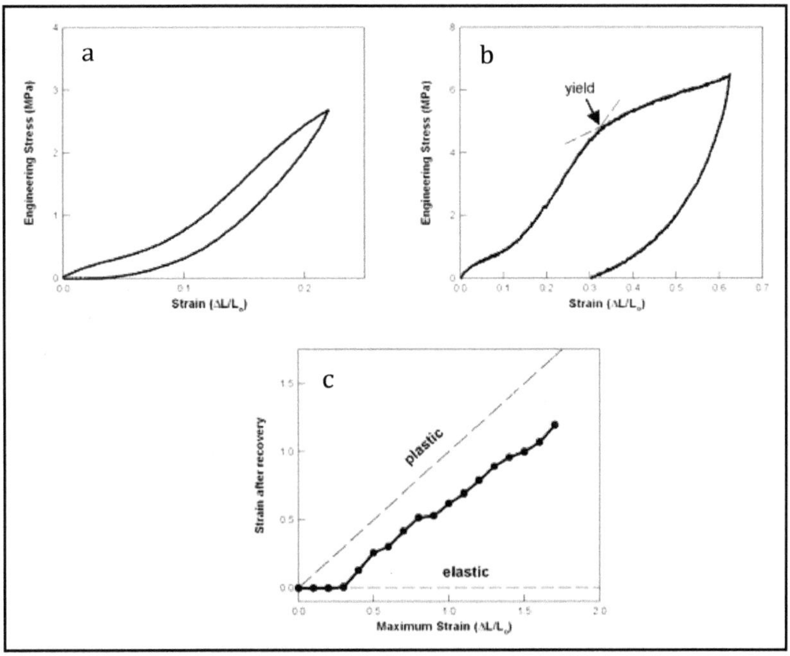

Figure 2.6 Recovery behaviour of slime threads in sea water.[11] (a) A typical load-unload cycle in region I, showing completely reversible deformation. (b) Typical load-unload cycle into region II, showing that deformation past the yield strain is mostly plastic. (c) Results from recovery trials in which threads were extended to a given strain, held, and allowed to recover. Note the dramatic change in recovery behaviour between strains of 0.3 and 0.4. With permission from *Biophysical Journal*.[11]

This transition from elastic to plastic behaviour likely correlates with the onset of a conformational change in the intermediate filament proteins that make up the slime threads. Although their initial properties are soft and elastic, slime threads have a complex stress–strain curve that includes a dramatic strain-hardening region. Rubber can fail under stresses as low as 2 MPa,[20] whereas hydrated slime threads fail at a stress of approximately 180 MPa (Fig. 2.5). This added strength likely arises from additional non-covalent cross-linking that comes about as a result of the protein conformational changes alluded to above, and likely involves the formation of amyloid-like structures.

2.5 Structural Considerations

The first indication that slime thread proteins adopt an amyloid like conformation came from Congo red staining of whole hagfish slime that had been mechanically deformed in some way (Fig. 2.4). Further experiments established that the onset of positive Congo red staining (i.e., metachromasia and birefringence) corresponds with the yield point of slime threads in water, or a tensile strain of approximately 35% strain (Fig. 2.7).

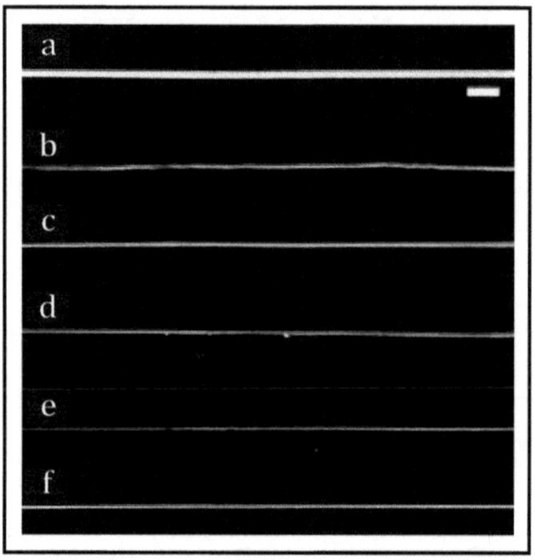

Figure 2.7 Congo red staining of slime threads stretched in sea water visualized with dark field polarized light microscopy.[11] (a) Unstrained and unstained threads are birefringent. Congo red staining of control threads and threads stretched to less than strains of 0.35 lose their birefringence and swell considerably (data not shown). (b–f) Threads stained with Congo red after extension to strains greater than 0.35 retained birefringence and mechanical integrity, and displayed increasing metachromasia as strain increased. Threads appeared orange-yellow when strained to 0.35 (b), green when strained to 0.50 (c), blue at 0.75 (d), blue-violet at 1.0 (e), and pale magenta to colourless at 1.50 (f). Scale bar is 10 μm. With permission from *Biophysical Journal*.[11] See also Colour Insert.

Curiously, the nature of the Congo red staining changes as a function of the strain history of the threads, with the resulting metachromasia becoming increasingly blue shifted with increasing strain. As far as we know, slime threads are the only system in which this kind of behaviour has been observed, and we currently do not have a firm grasp of its molecular underpinnings.

Further clues about the nature of the stretch-induced conformational transitions that occur in slime threads have come from wide angle synchrotron X-ray diffraction (Fig. 2.8).

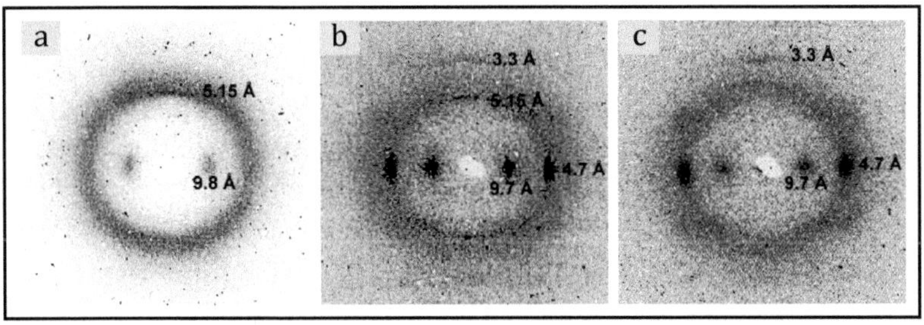

Figure 2.8 X-ray diffraction of hagfish threads strained in sea water.[11] (a) Unstrained threads exhibit a typical α pattern, whereas threads stretched to a strain of 1.0 exhibit a typical β pattern (c). Threads strained to 0.60 exhibit a mixed pattern (b). Diffraction maxima are labelled according to the molecular spacings (in angstroms) to which they correspond. With permission from *Biophysical Journal*.[11]

Native slime threads exhibit a pattern that is typical for α-keratins such as wool, with a meridional reflection at 5.15 Å, and an equatorial reflection at 9.8 Å.[11] These peaks correspond to the axial rise of coiled coils, and the spacing between adjacent coiled coils, respectively.[22,23] Threads stretched to a strain of 100% exhibit a typical "β-pattern", with strong equatorial reflections at 4.7 Å and 9.7 Å, and a meridional refection at 3.3 Å.[11] These peaks correspond to the spacings between protein chains within β-sheets, the spacings between adjacent β-sheets within β-sheet crystals, and the axial rise of β-sheets, respectively.[23,24] Bundles of threads stretched to a strain of 60% exhibit a mixed diffraction pattern, with all of the above reflections present. Together with the Congo red results, these data suggest that that the post-yield plastic deformation of slime threads corresponds to a stress-induced conversion of coiled coil α-helices into β-sheets that in turn stack into β-sheets crystals.[11]

These results point to some interesting differences between the structure of "conventional" amyloid fibrils and slime threads. One of the hallmarks of conventional amyloid is a cross-β structure, in which polypeptides are arranged into β-sheets that are oriented perpendicular to the long axis of the fibril.[25] In contrast, β-sheets in slime threads are arranged parallel to the long axis of the threads. This is not surprising given that β-sheet structures in slime threads arise from coiled coils that are parallel to the longitudinal axis of the thread. These coiled coils are transformed into β-sheets via tensile stresses that further increase the alignment of the proteins along the longitudinal axis.

Within coiled coil α-helices, hydrogen bonding occurs *within* the helices. When tensile stresses disrupt these intra-helix bonds in slime thread proteins, opportunities arise for the formation of inter-chain hydrogen bonding, in the form of β-sheets, which in turn can go on to stack into higher order structures like β-sheet crystals. It is this rise in inter-chain hydrogen bonding as well as the straightening of polypeptides that likely account for the dramatic strain hardening (i.e., increase in stiffness) that is seen in slime threads. In contrast, amyloid fibrils exhibit high initial stiffness due to extensive β-sheet bonding. Tensional stresses on amyloid fibres directly strain inter-chain β-sheet hydrogen bonds, and it is the gradual unzipping of these bonds that provides amyloid fibrils with high toughness (or energy to break). These bonds have also been described as "sacrificial" bonds that are required for longitudinal deformation, but do not lead to failure of the fibril because of the "hidden length" that is released when a row of hydrogen bonds has been broken.[13] In slime threads, sacrificial bonds and hidden length exist within coiled coil α-helices. When β-sheets are formed, because they are oriented parallel to the fibre axis, further straining of the material can only occur via the straining or breakage of peptide bonds, or the shearing of protein chains past one another, resulting in high stiffness of the material.

2.6 Implications for Other Materials Containing Intermediate Filaments

Although studies of the structure and mechanics of slime threads were originally undertaken to better understand the biophysics of hagfish slime, the results of this work have inspired novel research into mechanical properties of intermediate filaments in other systems. Since it was first recognized that 10 nm filaments within living cells are from the same gene family as the filaments that reinforce α-keratins (like wool), cell biologists have assumed that the mechanical properties of intermediate filaments in living cells can be approximated by the material properties of wool.[26] This assumption was seriously challenged, however, when it was shown that the bundle of intermediate filaments that make up slime threads are far more compliant and extensible than wool fibres.[11] This result raised two novel hypotheses—that intermediate filaments in cells are far softer and more extensible than previously assumed, and that the intermediate filaments in hard α-keratins like wool are modified in some way to make them stiffer and less extensible.

The first hypothesis regarding the mechanics of intermediate filaments in living cells makes a few critical predictions about how the intermediate network in living cells responds to mechanical stress. Specifically, if intermediate filaments in cells behave like slime threads, they should be able stretch to elastically to strains as high as 35%, and should resist breaking up to strains as high as 220%. Furthermore, strains greater than 35% should result in plastic deformation of the intermediate filament network. Testing these predictions requires subjecting cells to strains much higher than they typically experience in an experimental setting.

We are currently testing these predictions using a stable line of human keratinocytes that expresses a fluorescently tagged version of the intermediate filament gene keratin 14. Thus far, all of the data we have collected using a custom-built uniaxial cell stretching device is consistent with the predictions of this new model of intermediate filament mechanics in living cells.[27] Furthermore, we have found that cells subjected to these extreme degrees of uniaxial strain remain viable, which underscores the potential physiological relevance of the elasticity and high extensibility of intermediate filaments. These results also suggest the possibility that intermediate filaments could be transformed into amyloid-like structures in cells after extreme cellular deformations. The implications of this possibility will be discussed in the concluding section of this chapter.

Further evidence for this model has come from recent *in vitro* studies of the behaviour of single intermediate filaments probed using atomic force microscopy (AFM). Kreplak and colleagues have found that single intermediate filaments can be stretched with an AFM tip to strains as high as 250% before breaking, and they are currently in the process of attempting to derive the tensile stress–strain curve for a single intermediate filament using these same tools.[28] These researchers found that not only can keratin (Type I/II), desmin (Type III), and neuro filament (Type IV) intermediate filaments deform to 250% strain before breaking, they also found that the filaments did not return to their resting length after these kinds of deformations. This result is in perfect alignment with the post-yield plastic behaviour of hagfish slime threads.

This new model of intermediate filament mechanics in living cells raises the obvious question of why the properties of hard α-keratins are so different if they consist mostly of intermediate filaments. The main differences that need to be explained are the high elastic modulus of hard α-keratins like wool (approximately 2 GPa) compared to slime threads (approximately 6 MPa), the low yield strain of 2.5% (compared to 35% in slime threads), the high yield stress of approximately 40 MPa (compared to 3 MPa in slime threads), and the low extensibility of approximately 45% (compared to 220% in slime threads). One factor that could explain these differences is that the proteins within intermediate filaments in hard α-keratins are covalently cross-linked so that lines of force bypass the softer components that lend greater compliance to slime threads. We have hypothesized that these softer components are the terminal domains that flank the central α-helical rod domain,[11] but this has yet to be tested.

Another possibility is that the intermediate filaments in hard α-keratins are maintained in a semi-dehydrated state.[29] Support for this idea comes from the fact that slime threads tested in air have a stress–strain curve that is reminiscent of the curve for hard α-keratins tested in water. Specifically, dry slime threads exhibit high elastic modulus (in the several GPa range) and a low yield strain (2.4%) like hard α-keratins. The dehydration hypothesis is also supported by the fact that the yield stress of hard α-keratins tested in water is considerably higher than one would predict from cross-linking alone. One possible mechanism by which hard α-keratins could be maintained in a dehydrated state is via initial air drying, after which point they could be squeezed down upon by the polymerization of the keratin "matrix", which consists of a highly cross-linked network of proteins that surround the intermediate filaments.[29]

Another major difference between slime thread mechanics and hard α-keratins like wool is that post-yield deformations in the former lead to the conversion of α-helical secondary structure into amyloid-like β-sheets. In contrast, post-yield deformations (i.e.,

>2.5% strain) in wool are completely reversible (at least under physiological conditions), and do not lead to the formation of stable β-sheets. Interestingly, these deformations lead to the disruption of α-helical secondary structure, but somehow the formation of β-sheets is inhibited.[23] Although the mechanism is not fully known, the matrix proteins surrounding the intermediate filaments in wool are likely responsible. The matrix is believed to form an elastomeric network that acts in parallel to the intermediate filaments, and thus may provide a restoring force that inhibits formation of β-sheets and returns disrupted regions to their former α-helical state. If intermediate filaments in hard α-keratins are in fact maintained in a dehydrated state, even in water, this too could contribute to the reluctance of hard α-keratins to undergo a strain-induced α-to-β transition. Textile scientists have known for decades that wool fibres can be persuaded to undergo an α-to-β transition by holding them in a stretched state in the presence of steam.[30] This treatment likely disrupts the matrix and allows water to penetrate the intermediate filaments, which may lower the activation energy of the α-to-β transition.

While mammalian hard α-keratin epidermal appendages are reinforced with intermediate filaments that are dominated by α-helix, birds and reptiles evolved a unique hard keratin material that is reinforced with a completely different fibrous component dominated by β-sheet.[31] These so-called β-keratins make up scales, feathers, and beaks in these groups. Because the fibrils within β-keratins consist of proteins in β-sheets with the chains aligned primarily with the fibril axis, these materials exhibit high stiffness and do not exhibit a dramatic yield point like hard α-keratins do when the hidden length within α-helices is released.

2.7 Conclusions and Outlook

Hagfish slime threads are a case of a structural biomaterial that adopts amyloid-like properties. The fact that slime threads only take on an amyloid-like structure outside of the animal's body, and only after they have been draw-transformed by an external force, underscores the paucity of amyloid structures found within cells and within organisms. Slime threads are essentially bundles of intermediate filaments, and as such, they have proven to be a valuable model for understanding the mechanical properties of intermediate filaments in a variety of contexts, including the cytoskeleton of living cells, as well as the fibre-reinforced composites known as α-keratins. Studies of slime thread mechanics have also raised the intriguing possibility that intermediate filaments in living cells may adopt an amyloid-like structure under exceptional cases of mechanical stress. If this is true, intermediate filaments may be important for understanding the pathophysiology of conformational diseases, especially those for which mechanical trauma is a significant risk factor, such as Alzheimer's disease.

References

1. Fernholm, B., and Quattrini, A. (2008) *Copeia*, **1**, 126.
2. Martini, F. H. (1998) The ecology of hagfishes, in *The Biology of Hagfishes* (ed. Jorgensen, J. M, Lomholt, J. P., Weber, R. E., *et al.*), Chapman and Hall, New York, pp. 57–77.
3. Smith, C. R. (1985) Deep-sea research Part A. *Oceanographic Research Papers*, **32**(4), 417.
4. Jensen, D. (1966) *Scientific American*, **214**(2), 82.
5. Fudge, D. S., Levy, N., Chiu, S., and Gosline, J. M. (2005) *J. Exp. Bio.*, **208**, 4613.
6. Lim, J., Fudge, D. S., Levy, N., and Gosline, J. M. (2006) *J. Exp. Bio.*, **209**, 702–710.
7. Spitzer, R. H., and Koch, E. A. (1998) The hagfish skin and slime glands, in *The Biology of Hagfishes* (ed. Jorgensen, J. M, Lomholt, J. P., Weber, R. E., *et al.*), Chaman and Hall, London, pp. 109.
8. Downing, S. W., Spitzer, R. H., Salo, W. L., Downing, J. S., Saidel, L. J., and Koch, E. A. (1981) *Science,* **212**, 326.
9. Luchtel, D. L., Martin, A. W., and Deyrup-Olson, I. (1991) *Tissue Cell,* **23**, 939.
10. Koch, E. A., Spitzer, R. H., Pithawalla, R. B., and Downing, S. W. (1991) *Cell Tissue Res.,* **264**, 79.
11. Fudge, D. S., Gardner, K. H., Forsyth, V. T., Riekel, C., and Gosline, J. M. (2003) *Biophysical J.,* **85**(3), 2015.
12. Slotta, U., Hess, S., Spiess, K., Stromer, T., Serpell, L., and Scheibel, T. (2007) *Macromol. Biosci.,* **7**(2), 183.
13. Mostaert, A. S., and Jarvis, S. P. (2007) *Nanotechnology,* **18**, 044010.

14. Koch, E. A., Spitzer, R. H., Pithawalla, R. B., Castillos, F. A., III, and Parry, D. A. (1995) *Int. J. Biol. Macromol.,* **17**(5), 283; Koch, E. A., Spitzer, R. H., Pithawall, R. B., and Perry, D. A. (1994) *J. Cell Sci.,* **107**, 3133.
15. Lazarides, E. (1980) *Nature,* **283**(5744), 249.
16. Parry, D. A., and Steinert, P. M. (1999) *Quart. Rev. Biophys.,* **32**(2), 99.
17. Fraser, R. D., MacRae, T. P., and Rogers, G. E. (1972) *Keratins: Their Composition, Structure, and Biosynthesis,* Charles C. Thomas, Springfield, IL.
18. Herrmann, H., Hesse, M., Reichenzeller. M., Abei, U., and Magin, T. M. (2003) *Int. Rev. Cytol.,* **223**, 83; Guérette, D., Khan, P. A., Savard, P. E., and Vincent, M. (2007) *BMC Evol. Biol.,* **7**, 26–35.
19. Kreplak, L., and Fudge, D. (2007) *Bioessays,* **29**, 26.
20. Lillie, M. A., and Gosline, J. M. (2002) *Int. J. Biol. Macromol.,* **30**, 119.
21. Shadwick, R. E., and Gosline, J. M. (1983) *Can. J. Zool.,* **6**(8), 1866.
22. Pauling, L., and Corey, R. B. (1953) *Nature,* **171**, 59.
23. Kreplak, L., Doucet, J., and Briki, F. (2001) *Biopolymers,* **58**(5), 526.
24. Bendit, E. G. (1960) *Text. Res. J.,* **30**(8), 547.
25. Bonar, L., Cohen, A. S., and Skinner, M. M. (1969) *Proc. Soc. Exp. Biol. Med.,* **131** (4), 1373.
26. Mofrad, R. K., and Kamm, R. D. (2006) *Cytoskeletal Mechanics: Models and Measurements,* Cambridge University Press, Cambridge, UK; Bray, D. (2001) *Cell Movements: From Molecules to Motility,* 2nd edn, Garland, New York; Howard, J. (2001) *Mechanics of Motor Proteins and the Cytoskeleton,* Sinauer Associates, Sunderland, Massachusetts.
27. Fudge, D., Russell, D., Beriault, D., Noore, W., Lane, E. B., and Vogl, A. W. (2008) *PLoS One,* **3**(6), e2327.
28. Kreplak, L., Bar, H., Leterrier, J. F., Herrmann, H., and Aebi., U. (2005) *J. Mol. Biol.,* **354**(3), 569; Kreplak, L., Herrmann, H., and Aebi, U. (2008) *Biophysical J.,* **94**(7), 2790–2799,.
29. Fudge, D. S., and Gosline, J. M. (2004) *Proc. R. Soc. London, Ser B,* **271**(1536), 291.
30. Feughelman, M. (1968) *Text. Res. J.,* **38**, 1136.
31. Astbury, W. T., and Marwick, T. C. (1932) *Nature,* **130**, 309.

Chapter 3

Fungal Fibrils: Application of the Amyloid Polymer Structure by Fungi

Margaret Sunde,[a] Matthew D. Templeton,[b] and Ann H. Kwan[a]

[a]*School of Molecular Bioscience, University of Sydney, Sydney 2006, Australia*

[b]*The Horticultural and Food Research Institute of New Zealand, Mt Albert Research Centre, Auckland, New Zealand*

m.sunde@mmb.usyd.edu.au

Hydrophobins are low molecular weight proteins found ubiquitously in filamentous fungi. They are secreted as soluble monomers, but polymerize spontaneously at hydrophobic/hydrophilic interfaces to coat aerial structures such as conidia and fruiting bodies. Hydrophobins play multiple roles in fungal biology, ranging from conferring water resistance to fungal structures, to assisting the formation of aerial hyphae by reducing the surface tension of the growth medium, as well as mediating interactions between the fungus and its environment or host. Many hydrophobins polymerize into fibrillar assemblies (known as rodlets) that closely resemble amyloid fibrils, both in terms of structure and physical properties. In addition, the rodlets assemble laterally into monolayers that are amphipathic. Hydrophobin rodlets therefore represent an application of the amyloid structural scaffold by nature to serve

The Functional Fold: Amyloid Structures in Nature
Edited by Suzi Jarvis and Anika Mostaert

ISBN 978-981-4267-40-3 (Hardcover), 978-981-4267-62-5 (eBook)
www.panstanford.com

specific biological purposes, in contrast to the amyloid deposits that are commonly associated with disease states and protein misfolding. This chapter will focus on what is currently known about the sequence, structure, assembly mechanism, and physical properties of hydrophobin rodlets and discuss the diverse functions played by these fibrils in fungal biology and development.

3.1 Introduction

Fungi are heterotrophic organisms that play a key role in nutrient recycling in the ecosystem. As such, fungi generally grow in moist environments such as leaf litter, wood, and living plants, and less commonly in animal tissue through the apical growth of hyphae.[1,2] Once nutrients have been depleted from the immediate environment, and in order to colonize new territory, fungi produce aerial hyphae and reproductive structures such as conidiophores and fruiting bodies (e.g., mushrooms or brackets) to enable spore dispersal. The transition from the moist hydrophilic environment to the hydrophobic conditions of the air is essential for the effective dispersal of spores. Filamentous fungi are able to transverse the boundary between dry and moist environments successfully due to the presence of a unique family of proteins, called hydrophobins. Many of these fungal hydrophobins self-assemble into a functional amyloid structure that forms an amphipathic layer at critical hydrophilic/hydrophobic interfaces, allowing fungi to successfully colonise a range of environments.

3.2 Early Observations of Rodlets

A regular, fibrillar structure was first observed 40 years ago on the outer surface of freeze-etched spores from mould fungi such as *Penicillium*,[3,4] *Aspergillus*,[5,6] *Oidiodendron truncatum*,[7] and *Geotrichum candidum*.[7] These fibrillar structures were given the name "rodlets". Micrographs showed that rodlets ranged from approximately 100 to 18,000 Å in length and were usually grouped in bundles of approximately 300 to 2500 Å in width (Fig. 3.1a,b).[5] Beever and colleagues showed that the rodlets from *Neurospora crassa* formed a monolayer on the spore surface and were not interwoven or overlaid.[8–10] The first clue as to their function came from the observation by Dempsey and Beever that spores from a particular *N. crassa* mutant lacked the characteristic rodlet pattern on the surface (Fig. 3.1c).[11] This mutant had been described by Seletrennikoff as having an "easily wettable phenotype", that is, the aerial hyphal mass of this mutant, growing on agar absorbed water, whereas a water droplet would remain intact and suspended on wild-type hyphae.[12] In addition, the mutant could liberate free conidia only into water, but not into air, as the wild-type fungus could. In 1978, Dempsey and Beever proposed that the function of rodlets was to impart water repellency and to allow spores to be easily dispersed in air.[8] The rodlets were initially thought to be composed of biopolymers such as cutin or sporopollenin.[3,4] Further analysis of the rodlet layer was complicated by the fact that the rodlets were insoluble in organic solvents, hot detergent, and alkaline solutions. Nonetheless, it became clear that the major component of the rodlets was protein but the insolubility of the rodlets prevented detailed chemical analysis.[13]

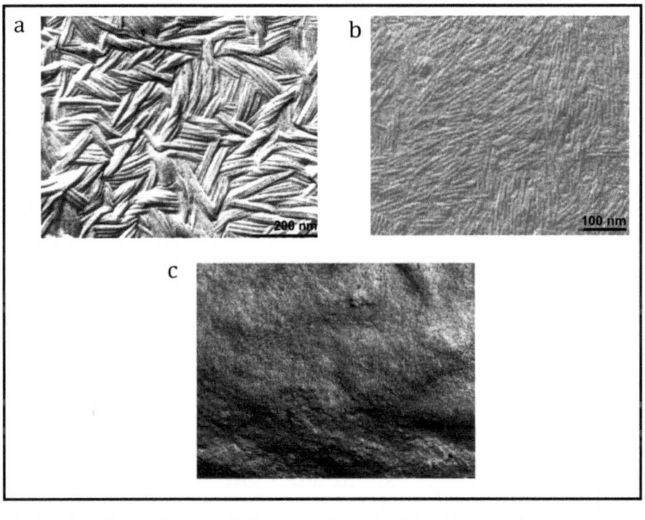

Figure 3.1 Rodlets from different fungi. (a) Freeze-fracture image of *Neurospora crassa* spores, showing rodlets composed of the hydrophobin EAS. (b) Freeze-fracturing and shadowing show the hydrophobin SC3 assembled into rodlets on the surface of aerial hyphae. (c) The surface of a *N. crassa* spore from a knock-out strain that does not express the EAS protein is smooth, lacks rodlets, and demonstrates the "easily wettable" phenotype. Part b reproduced with permission from Fig. 3 of Wessels, J. G. H. (1996). Copyright Elsevier Ltd.

A different line of investigation led to the identification of hydrophobins as the putative product of genes that are abundantly expressed at certain stages in fungal development.[14] In 1984, Wessels and colleagues reported high levels of novel mRNAs expressed during the growth of aerial hyphae and fruiting bodies in *Schizophyllum commune.* These mRNAs encoded a family of low molecular weight, hydrophobic proteins which they termed "hydrophobins".[15] The realization that rodlets were composed of hydrophobins came when it was demonstrated that solubilization of the rodlet layers from the cell walls of aerial hyphae or fruiting bodies by treatment with formic or trifluoroacetic acid generated the monomeric forms of the hydrophobins SC3 and SC4, respectively. The discovery that rodlet layers could be solubilized by these acids paved the way for the analysis of rodlet proteins from a number of fungi. It was also found that the hydrophobin proteins were secreted into the culture medium in a monomeric form, but that agitation of the solution rapidly converted the protein into a sodium dodecyl sulfate (SDS)-insoluble aggregate.[16] Soon after, the hydrophobin EAS (protein encoded by the easily wettable gene) from *N. crassa*[17] and RodA, the hydrophobin protein from *Aspergillus nidulans,*[18] were cloned and shown to encode proteins that were clearly homologous to the hydrophobins of *S. commune.* It was subsequently shown that disruption of the former gave rise to the "easily wettable phenotype" in *N. crassa* (Fig. 3.1c) that had been described by Seletrennikoff earlier,[19] and disruption of RodA in *A. nidulans* gave rise to non-hydrophobic spores that lacked rodlets.[18] Furthermore, the purified *N. crassa* rodlet layer, solubilized by trifluoroacetic acid, was shown to be the proteolytically cleaved product of *eas*. In addition, sequence comparison identified putative hydrophobin proteins in several other organisms, including the insect pathogenic fungus *Metharizium anisopliae,*[20] the rice blast fungus *Magnaporthe grisea,*[21] and *Ophiostomi ulmi*, the causative agent of Dutch elm disease.[22] A recent search of the NCBI nr database found 128 different hydrophobin sequences.[23] Whole genome sequencing projects have now revealed that hydrophobins exist in fungi as small gene families of between 2 (*N. crassa*) and 24 (*Coprinus cinereus*) members (http://www.broad.mit.edu/annotation/fgi/).

3.3 The Two Classes of Hydrophobins

Comparative analysis of hydrophobin sequences showed that in spite of the low levels of sequence homology between these proteins, they share a very distinctive arrangement of eight cysteine residues in a unique pattern (X_{2-38}-C-X_{5-10}-CC-X_{11-44}-C-X_{8-23}-C-X_{5-9}-CC-X_{6-21}-C-X_{2-13}), including two pairs of adjacent cysteine residues (Fig. 3.2).[24] Within the hydrophobin family, proteins can be further separated into Class I and Class II hydrophobins, on the basis of the distinct hydropathy profiles displayed by these classes and the conservation in the spacing of amino acids between the cysteine residues. This distinction between the sequence characteristics of Class I and Class II hydrophobins has been maintained and clarified as increasing numbers of hydrophobins have been sequenced.[25]

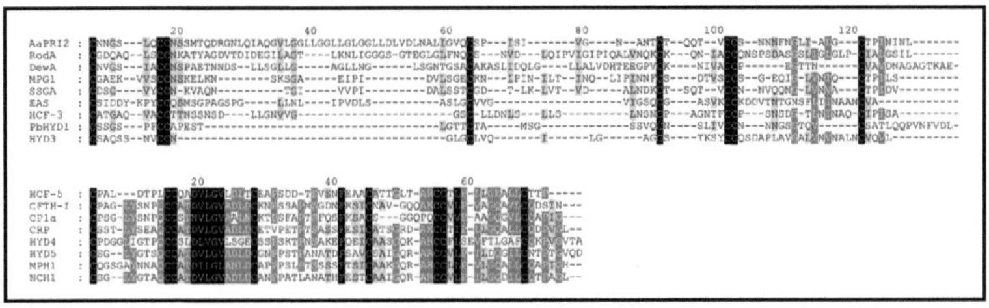

Figure 3.2 Alignment of Class I and Class II hydrophobins. Alignment of representative set of Class I and Class II hydrophobin sequences. This illustrates the conserved pattern of eight cysteine residues, including the two pairs of adjacent cysteines, and the differences in distribution of these cysteine residues between the two classes. It is also clear that the Class I sequences display very limited sequence homology and great variation in the length of the inter-cysteine regions. In contrast, the Class II hydrophobins display much higher conservation of sequence and loop length. Extent of sequence conservation is indicated by shading from black (absolutely conserved), through shades of grey, to white (not conserved).

The monolayers formed by members of both classes of hydrophobin are amphipathic. When the layers are assembled *in vitro*, their orientation depends on the nature of the surface, that is, hydrophobin monolayers are able to reverse the wettability of surfaces.[26,27] When assembled on a glass slide, SC3 raises the water contact angle from approximately 40° for naked glass to 55°, whereas an SC3 coating on Teflon™ reduces the water contact angle[28] from approximately 110° to 32°. A similar effect can be observed for EAS monolayers (Fig. 3.3). Both Class I and Class II hydrophobins are extremely surface active, that is, they are able to reduce the surface tension of solutions to which they are added. For example, SC3 from *S. commune*, and the Class II hydrophobin cryparin from the causal agent of chestnut blight, *Cryphonectria parasitica*, are able to lower the water surface tension[26] from 72 to 32 mJ/m^2. The ability of hydrophobins to reduce water surface tension is thought to play a functional role in some fungal systems, where hydrophobins are secreted into the growth medium and allow the hyphae to grow up into the air under reduced surface tension.[29]

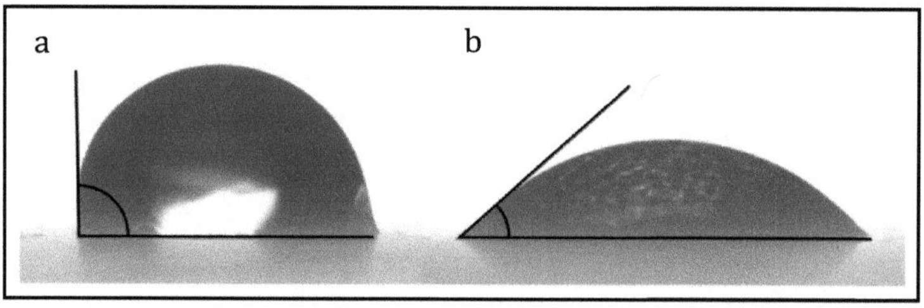

Figure 3.3 Hydrophobins can reverse the wettability of surfaces. Photograph showing water droplets sitting on (a) a Teflon™-coated slide and (b) EAS-coated Teflon™ slide. The water contact angle is marked and can be seen to be significantly reduced in (b) when compared to (a).

Despite the many similarities that exist between Class I and Class II hydrophobins, the bioinformatics-based division of hydrophobins mirrors key physiochemical differences between the two classes. Notably, class II hydrophobins form assemblies without a regular morphology and can be solubilized in alcohol solutions and in hot SDS. So far, all of the hydrophobins that form the robust, regular rodlets identified on outer fungal surfaces are strictly members from the Class I family. These rodlets are fibrillar in morphology, insoluble in SDS, and can only be solubilized by strong acid treatment, all of which are characteristics which call to mind amyloid-like structures.

3.4 Hydrophobins and Amyloid

The similarities between the morphological and structural characteristics of Class I hydrophobin rodlets and amyloid fibrils were first noted by Wosten and de Vocht.[30] To date, SC3 and EAS rodlets have been shown to exhibit some of the defining characteristics of amyloid (Fig. 3.4a).[31–33] Both types of protein assemblies appear long,

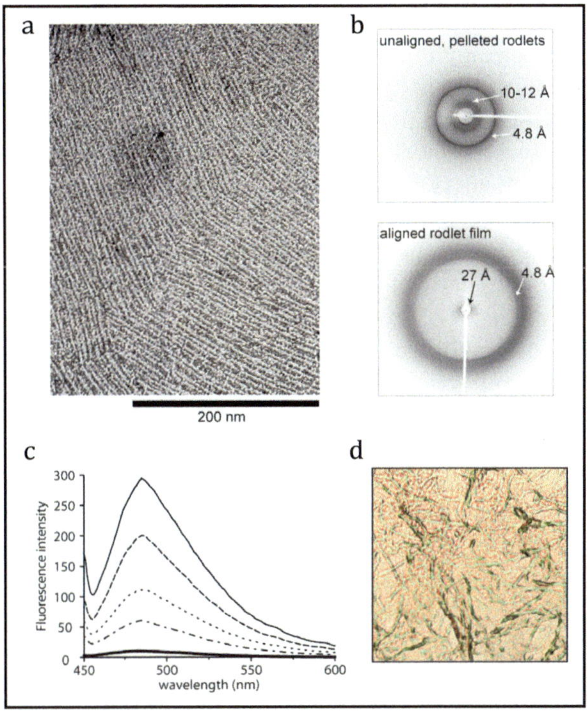

Figure 3.4 Hydrophobin rodlets display many features characteristic of amyloid fibres. (a) Transmission electron micrograph showing long, straight, and unbranched EAS rodlets with a width of approximately 10 nm. (b) X-ray fibre diffraction image of unaligned, pelleted rodlets displays reflections at 4.8 and 10–12 Å, while aligned EAS rodlet layer displays an additional reflection at approximately 27 Å, but no reflection at 10–12 Å. Reflections at 4.8 and 10–12 Å are highly characteristic of spacing between β-strands and β-sheet stacks, respectively. (c) Only EAS solutions which have been vortexed to form rodlets can bind Thioflavin-T. Concentrations of EAS solution used are as follows: 0.3 mM (thin black line), 0.2 mM (dash line), 0.1 mM (dotted line), and 0.05 mM (dotted and dash line). In contrast, unvortexed EAS solution shows no binding, even at 0.3 mM (thick black line). (d) Gold-green birefringence can be observed when EAS rodlets are stained with the amyloid-specific dye Congo red, viewed between cross-polarizers. See also Colour Insert.

straight, and unbranched in the electron microscope. The rodlets are 50–120 Å wide and appear to be composed of protofilaments that are approximately 2.5 Å wide. The rodlets form monolayers that are approximately 70–100 Å thick.[10,33,34] Given the sequence and physicochemical similarities of the Class I hydrophobins, it seems likely that these amyloid-like features will be common to all rodlets formed by this group of hydrophobins.

Preparations of unaligned EAS rodlets (produced by vortexing and then harvested by centrifugation) display a cross-β X-ray fibre diffraction pattern (Fig. 3.4b).[35] The dominant reflections in the pattern are at 4.8 and 10–12 Å, consistent with the inter-strand and inter-sheet spacings present in the β-sheet structure. The X-ray diffraction pattern obtained from a monolayer rodlet film formed by drying EAS rodlets in a magnetic field displays some preferential orientation of the inter-strand spacing and an additional perpendicular reflection at approximately 27 Å, which matches the diameter of the monomer. These reflections are consistent with the hydrogen bonding between strands lying in the direction of the rodlet long axis. There is no prominent inter-sheet spacing in this pattern, probably because this is a monolayer sample and so the inter-sheet spacing is not repeated in this direction.

Similar to amyloid fibrils, Class I hydrophobin rodlets display yellow-green birefringence when stained with the dye Congo red and viewed between cross-polarizers; they also bind to Thioflavin-T, and the dye exhibits enhanced fluorescence when bound to rodlets (Fig. 3.4c,d).[31–33,36] The monomeric, unassembled forms of the hydrophobins do not bind to these dyes, indicating that a structural change must occur upon rodlet formation to generate the extended, stacked β-sheet structure. Taken together, the nature of the X-ray fibre diffraction pattern and the specific interactions with Congo red and Thioflavin-T suggest that Class I hydrophobin rodlets have an underlying ordered β-sheet core.

Amyloid fibrils are composed of different numbers of sub-filaments. The number and arrangement of the sub-filaments appears to differ depending on the nature of the polypeptide forming the fibril and also the conditions under which fibrillogenesis occurs. Twisting of fibrils is also often observed. However, amyloid fibrils do not generally associate laterally in a specific or organized fashion over the length of the fibril. Rodlets, by contrast, show specific and directed lateral assembly and generate monolayers that are amphipathic. This implies that, in addition to the forces that drive longitudinal intermolecular assembly to result in amyloid-like rodlets, there is another dimension of molecular recognition and organization that drives lateral association of rodlets and orients them with respect to the surface upon which assembly takes place. On fungal structures, the hydrophobins assemble in a way that presents the hydrophobic face to the outside of the aerial structure, while the internal surface, adjacent to the cell wall, is hydrophilic.[10] A recent study of *Aspergillus fumigatus* spore germination has demonstrated this with clarity, showing the disappearance in spore surface hydrophobicity as germination occurs (Fig. 3.5).[37]

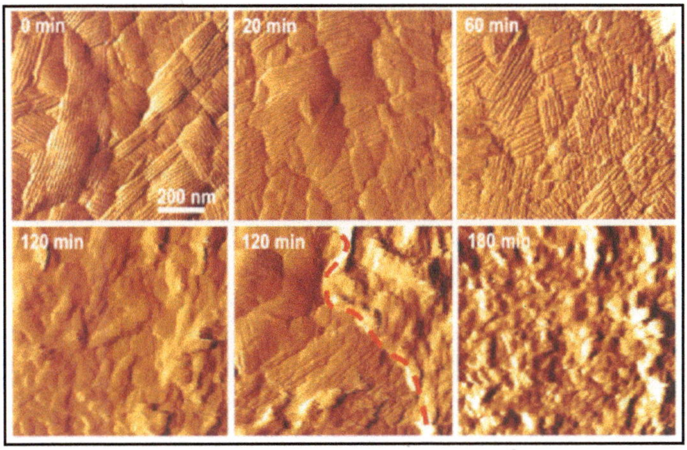

Figure 3.5 Structural dynamics of single germinating conidia. Series of high-resolution AFM deflection images recorded on a single spore during germination. Within 3 hours the crystalline rodlet layer changes into a layer of amorphous material, presumably reflecting inner cell wall polysaccharides. After two hours, both rodlet and amorphous regions were found to coexist (lower middle panel, left and right, respectively, of dashed line). Figure reproduced with permission from Dague *et al.* (2007). Copyright Biophysical Society. See also Colour Insert.

In contrast to the amyloid-like rodlets, the monolayers of Class II hydrophobins are not fibrillar, or as robust, and do not bind Thioflavin-T. Although they show some regular ordering, assemblies of Class II hydrophobins do not display any reflections in either X-ray scattering or diffraction patterns that could arise from β-sheet structure.[38,39] Therefore, although both classes of hydrophobins form amphipathic monolayers at surfaces, only the Class I hydrophobins do so on the scaffold of an amyloid-like structure.

3.5 Structural Change During Rodlet Formation

The three-dimensional structures of the monomeric, unassembled forms of two Class II and one Class I hydrophobins have been determined. The structures of the two highly homologous Class II hydrophobins from *Trichoderma reesei*, HFBI and HFBII, have been determined by X-ray crystallography.[40-43] The solution structure of EAS, the Class I hydrophobin, from *N. crassa* has been determined by triple-resonance NMR methods.[35] Remarkably, given the low levels of sequence similarity between the two classes of hydrophobins and the fact that the assemblies formed by these hydrophobins are different in morphology and stability, EAS and the HFBI/II have the same basic topology in the unassembled, monomeric form (Fig. 3.6a–c). This fold is unique to the hydrophobin family of proteins. The structures of HFBI and HFBII consist of a core β-barrel comprised of two adjoining β-hairpins, with an α-helix linked to the

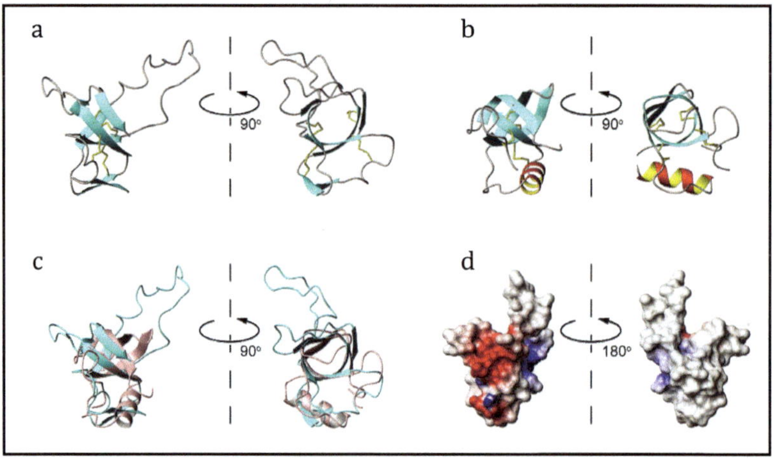

Figure 3.6 The three-dimensional structures of a Class I and a Class II hydrophobin. (a) Ribbon diagrams of the Class I hydrophobin EAS (PDB code 2FMC) and (b) the Class II hydrophobin HFBII (PDB code 1R2M). Cysteine side-chains are shown in yellow. (c) A superposition of EAS (cyan) and HFBII (pink) showing the β-barrel core is highly similar. (d) Electrostatic surface of EAS. Negatively and positively charged residues are illustrated in red and blue, respectively. The charges can be seen to cluster on a single face of the structure, while the diametrically opposing face is hydrophobic. See also Colour Insert.

outside of the barrel via a disulphide bond. Two disulphide bonds cross-link the two strands within each of the two β-hairpins and the fourth disulphide bond connects the N-terminal loop to the core β-barrel.

Similarly, the core structure of EAS is centred on two interlocking β-hairpins that form a four-stranded β-barrel. However, in place of the α-helix found in the HFBI/II crystal structures, there is a short two-stranded antiparallel β-sheet. The other significant difference is that EAS has two disordered loops that extend from the barrel. The longer of the two lies between the third and fourth cysteines. This is a region of high variation between Class I hydrophobins, both in terms of sequence and length. This disordered loop is absent from Class II hydrophobins. The structure of EAS demonstrates how the inter-cysteine variation observed in hydrophobins can be accommodated on the periphery of the conserved hydrophobin barrel and it is likely that this is the case for all Class I proteins, given they all share the same pattern of cysteine residues.

The three-dimensional structures of these hydrophobins reveal the structural basis for their surface activity and ability to form amphipathic monolayers. Both Class I and Class II molecules display large hydrophobic patches on the surface, which account for at least one-fifth of their total surface area. In EAS, there are relatively few charged residues and these are clustered on a single face of the protein, giving rise to a molecule with distinct, opposing hydrophilic and hydrophobic faces (Fig. 3.6d). The charged residues in HFBI and HFBII are more evenly distributed over the surface so that the molecules do not display such a striking amphipathic character. However, the presence of the hydrophobic regions on the surface of these proteins is thought to drive the orientation of both classes of hydrophobin at a hydrophobic/hydrophilic interface and thus, the assembly of the monolayer.

A comparison of the monomeric structures of the Class I and Class II members does not obviously identify features which can explain the fundamental differences in the secondary structure content, stability, and morphology of their assembled forms. Instead, it is possible that the differences between the classes of hydrophobins arise from variations in their conformational plasticity and dynamics, and in their differing ability to access structural intermediates which are prone to intermolecular aggregation through hydrogen bonding. In the amyloid field, for example, amyloidogenic variants of proteins

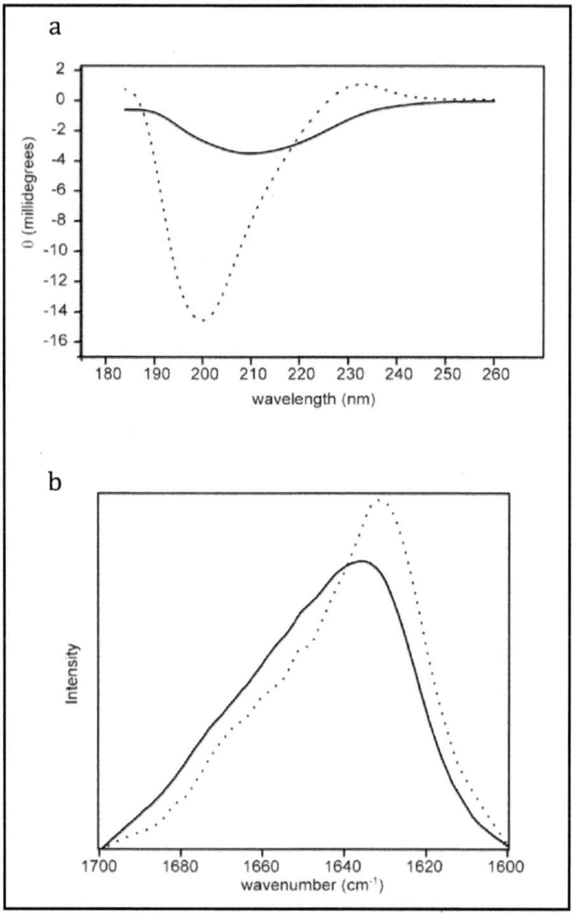

Figure 3.7 Hydrophobin rodlets experience a significant increase in β-sheet structure upon assembly. (a) Circular dichroism spectra of EAS in water (dotted line) and dried onto a quartz cuvette (solid line). The minimum shifts from 200 to approximately 210 nm upon rodlet formation indicating the presence of β-sheet structure. (b) Attenuated total reflectance Fourier transform infrared spectra of SC3 under two conditions: non-assembled SC3 deposited onto a germanium plate (dark thick line) and vortexed SC3 on a germanium plate (dotted line). A shift to the lower wavelengths of approximately 1630 nm (indicative of the presence of mainly β-sheet structure) can be seen in the vortexed sample. Vertical scale is arbitrary. Reprinted with permission from de Vocht *et al.* (1998). Copyright 1998 Biophysical Society.

are often those that display enhanced conformational dynamics and which populate partially structured intermediates.[44] Like many amyloid-forming proteins, the Class I hydrophobins appear to undergo some level of conformational change when rodlets form, whereas the Class II hydrophobins do not.[45]

Techniques such as circular dichroism spectropolarimetry (CD) and Fourier transform infrared spectroscopy (FTIR) have been used to follow the structural changes which occur during rodlet formation by Class I hydrophobins. The CD spectrum of soluble SC3 indicates that the protein has elements of β-sheet structure, together with some random coil. When colloidal Teflon™ is added to a solution of SC3, spectral features indicative of α-helical structure appear. These α-helical elements may be relevant to the polymerization process in SC3 since an on-pathway intermediate has been proposed to be important in rodlet formation.[46–48] The CD spectra from the soluble forms of other Class I hydrophobins including EAS, SC4, and ABH3 indicate that they all contain varying amounts of β-sheet and random coil structure.[30,33,49] However, in all cases, a significant increase in β-sheet structure is detected when rodlets form, which is also commonly observed upon amyloid formation (Fig. 3.7).

The FTIR spectrum of soluble SC3, dried onto a germanium plate, is consistent with the presence of a mixture of secondary structure[46]. Deconvolution and curve fitting analysis of the single major amide I band at 1636 cm^{-1} suggests that soluble SC3 is composed of 23% α-helix, 41% β-sheet, and 16% β-turn, with the remaining 20% being random coil. Drying down of SC3 onto a silanized germanium plate, which increases the surface hydrophobicity, changed the profile to one with mainly α-helix and β-sheet contributions, suggesting that the assembling SC3 intermediates might contain helical structure. The final assembled rodlet form of SC3 is β-sheet rich,[46,47] as is the rodlet form of EAS.[33] Polarization-modulated infrared reflectance adsorption spectroscopy (PM-IRRAS) indicates that SC3 localizes to the surface of the solution rapidly, where it undergoes a conformational change to a mainly β-sheet form.[47] This form of spectroscopy is also able to report on the orientation of secondary structure elements relative to the incident surface. In the case of SC3 rodlets, the nature of the amide signal indicates that hydrogen bonds are oriented preferentially parallel to the air–water interface. This shows that the plane of the β-sheet structure, in which the hydrogen bonds lie, must be co-planar with the long axis of the rodlets. Other studies of SC3 oligomerization, which have been monitored by both EM and PM-IRRAS,[47] suggest that the hydrophobin must localize to the surface and reach a critical concentration, in order to initiate the structural conversion to the mainly β-sheet form and for rodlets to form.

3.6 Specialization and Functional Significance of These Functional Amyloids

Hydrophobins have been identified in three of the major fungi: Ascomycota, Basidiomycota, and Zygomycota. These include plant and human pathogens, edible mushrooms, and fungi that form symbiotic relationships with cyanobacteria (lichens) and plant roots (mycorrhiza).[50] In lichens, hydrophobins line the air pockets preventing waterlogging and facilitating gas exchange. Hydrophobins are highly expressed during the formation of mycorrhizae, but a role for these proteins has not yet been determined.[51] Indeed the potential role of hydrophobins in soil ecology has only recently been appreciated.[52] This wide distribution and the many different functions of hydrophobins are indicative of the utility of these proteins and shows how important bridging the interfaces between the air, solid surfaces, and liquids is to the fungal life cycle.

Some fungi appear to have evolved a different family of proteins that carry out some of the functions of hydrophobins. A recently discovered example occurs in *Ustilago maydis*, the fungus that causes maize smut. Although the genome of this fungus encodes two hydrophobins, it uses a repellent protein to generate a hydrophobic surface on aerial hyphae and to mediate attachment to hydrophobic surfaces.[53] In contrast, no hydrophobins or repellents have been identified in yeasts, even though several genomes have been sequenced, so hydrophobins may be dispensable for growth in generally aqueous habitats.

Multiple hydrophobins are expressed within individual fungi, and the expression of these genes is temporally and spatially regulated, suggesting specialization for different roles in development. More than twice the number of Class I hydrophobins have been discovered, compared to Class II; members and the latter are found only in ascomycetes, suggesting that they may have evolved convergently.[25,50]

This may explain why only the Class I hydrophobins represent an application of the amyloid polymer fold.

While all Class I hydrophobins form rodlets with similar morphologies and physical properties, they appear to play distinct functional roles (Fig. 3.8). For example, in *S. commune*, SC3 is able

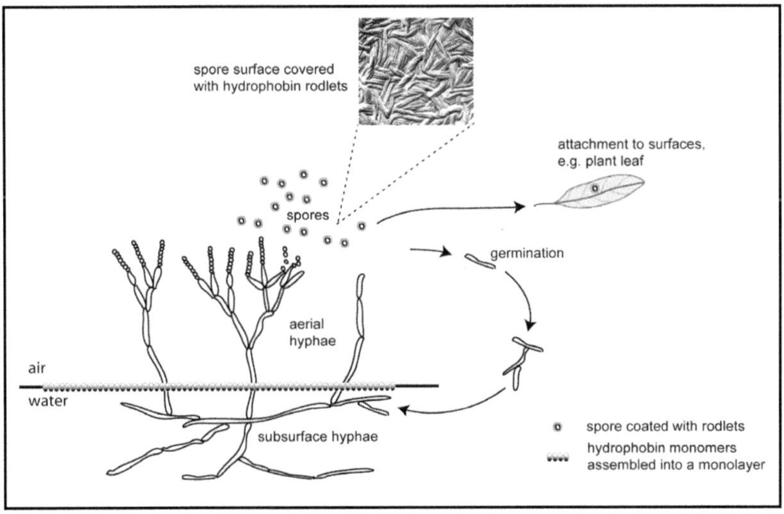

Figure 3.8 Role of hydrophobins in the fungal life cycle. Schematic illustration of the some of the roles played by hydrophobins in the life cycle of filamentous fungi. Hydrophobins are secreted into the moist growth environment and assemble into an amphipathic monolayer at the air–water interface. This acts to reduce surface tension, allowing the hyphae to breach the air water interface. Mature spores are also coated with a protective, amphipathic hydrophobin layer which resists wetting and facilitates the dispersal of the spores in air. In some systems, hydrophobin rodlets are involved in mediating the interaction between fungi and host surfaces.

to lower surface tension when secreted into the growth medium, coat aerial hyphae and mediate attachment to hydrophobic surfaces. On the other hand, the SC4 protein forms a hydrophobic layer that lines gas channels specifically within fruiting bodies.[54] Similarly, the human pathogen *A. fumigatus* expresses two highly related hydrophobins, RodA and RodB, but only RodA is responsible for forming a hydrophobic rodlet layer that coats the conidia. The RodA layer allows conidia dispersal and adhesion to host structures, and offers resistance to killing by host cells. In contrast, the hydrophobin RodB is thought to play a role in maintaining the structure of the conidial cell wall.[55,56] In *M. grisea*, the rice blast fungus, expression of the Class I hydrophobin MPG1 is necessary for formation of the rodlet layer on aerial hyphae and conidia, and is also required for normal appressoria formation and efficient infection and hence pathogenicity.[21] Interestingly, despite the functional distinctions between different Class I hydrophobins, complementation experiments showed that six of the seven Class I hydrophobin genes tested (including SC1 and SC4 from *S. commune*, RodA and DewA from *A. nidulans*, EAS from *N. crassa*, and SsgA from *M. anisopliae*), could at least partially fulfil the various roles played by MPG1. Therefore, although being very diverse in amino acid sequence and expression pattern, Class I hydrophobins constitute a closely related group of functional amyloid-like proteins with similar physical properties and underlying functions.[57]

3.7 Conclusions

Hydrophobins are ubiquitously found in filamentous fungi and are required for the dissemination of spores. The fungal hydrophobin rodlets represent a functional adaptation of the polymeric properties of amyloid. Both Class I and II hydrophobins are secreted by fungi in a soluble, monomeric form which then self-assembles into a highly insoluble and ordered state. However, only the Class I hydrophobin rodlets can be strictly classified as "functional amyloid" since their assemblies have a fibrillar morphology and display typical amyloid characteristics, such as binding to amyloid-specific dyes and giving rise to a cross-β X-ray fibre diffraction pattern. Apart from sharing significant structural and morphological similarities with amyloid fibrils, two features are unique to the hydrophobin assemblies. The first is that the polymerization process extends in both the longitudinal and lateral directions, resulting in the formation of a film rather than filaments. Secondly, hydrophobin monomers as well as the assemblies that they form are highly amphipathic. In fact, the amphipathicity of the hydrophobin molecules is thought to be a major driver of the assembly process, and the amphipathic character of the assemblies is thought to underlie most, if not all aspects of hydrophobin function in fungal biology. A number of bacterial proteins, such as the chaplin, curli, and tafi proteins from *Streptomyces*, *Escherichia coli*, and *Salmonella*, respectively, exhibit similar properties and serve similar biological functions to hydrophobins (Chapter 7). However, despite the visual resemblance of these bacterial protein assemblies to hydrophobin rodlets, there is little similarity in the amino acid sequences between these proteins. This use of the amyloid fibril scaffold in a number of micro-organisms represents a prime example of functional convergence found in nature.

Acknowledgements

This work was supported by Discovery and Linkage Project funding from the Australian Research Council (ARC; to M.S. and A.H.K.). M.S. is an R. D. Wright Fellow supported by the National Health and Medical Research Council (NHMRC) and A.H.K. is an Australian Postdoctoral fellow supported by the ARC.

References

1. Talbot, N. J. (1999) *Nature*, **398**, 295.
2. Elliot, M. A., and Talbot, N. J. (2004) *Curr. Opin. Microbiol.*, **7**(6), 594.
3. Sassen, M. M., Remsen, C. C., and Hess, W. M. (1967) *Protoplasma*, **64**, 75.
4. Hess, W. M., Sassen, M. M., and Remsen, C. C. (1968) *Mycologia*, **60**(2), 290.
5. Hess, W. M., and Stocks, D. L. (1969) *Mycologia*, **61**(3), 560.
6. Ghiorse, W., and Edwards, M. (1973) *Protoplasma*, **76**, 49.
7. Cole, G. T. (1973) *Can. J. Bot.*, **51**, 2413.
8. Beever, R. E., and Dempsey, G. (1978) *Nature*, **272**, 608.
9. Beever, R. E., Redgewell, R. J., and Dempsey, G. (1979) *J. Bacteriol.*, **140**, 1063.
10. Dempsey, G. P., and Beever, R. E. (1979) *J. Bacteriol.*, **140**, 1050.
11. Beever, R. E., and Dempsey, G. P. (1978) *Nature*, **272**(5654), 608.
12. Seletrennikoff, C. P. (1976) *Neurospora Newsletter*, **23**, 23.
13. Beever, R. E., Redgwell, R. J., and Dempsey, G. P. (1979) *J. Bacteriol.*, **140**(3), 1063.

14. Dons, J. J., Springer, J., de Vries, S. C., and Wessels, J. G. (1984) *J. Bacteriol.*, **157**(3), 802.
15. Wessels, J. G., de Vries, O. M., Asgeirsdottir, S. A., and Springer, J. (1991) *J. Gen. Microbiol.*, **137**(10), 2439.
16. Wosten, H. A., Asgeirsdottir, S. A., Krook, J. H., Drenth, J. H., and Wessels, J. G. (1994) *Eur. J. Cell Biol.*, **63**, 122.
17. Lauter, F. R., Russo, V. E., and Yanofsky, C. (1992) *Genes Dev.*, **6**(12A), 2373.
18. Stringer, M. A., Dean, R. A., Sewall, T. C., and Timberlake, W. E. (1991) *Genes Dev.*, **5**(7), 1161.
19. Bell-Pedersen, D., Dunlap, J. C., and Loros, J. J. (1992) *Genes Dev.*, **6**(12A), 2382.
20. St Leger, R. J., Staples, R. C., and Roberts, D. W. (1992) *Gene*, **120**, 119.
21. Talbot, N. J., Ebbole, D. J., and Hamer, J. E. (1993) *Plant Cell*, **5**(11), 1575.
22. Stringer, M. A., and Timberlake, W. E. (1993) *Plant Cell*, **5**(2), 145.
23. Yang, K., Deng, Y., Zhang, C., and Elasri, M. (2006) *BMC Bioinformatics*, **7**, Suppl 4, S16.
24. Wessels, J. G. H. (1994) *Ann. Rev. Phytopathol.*, **32**, 413.
25. Linder, M. B., Szilvay, G. R., Nakari-Setala, T., and Penttila, M. E. (2005) *FEMS Microbiol. Rev.*, **29**(5), 877.
26. Wosten, H. A., and de Vocht, M. L. (2000) *Biochim. Biophys. Acta*, **1469**(2), 79.
27. Wosten, H. A. B., Schuren, F. H. J., and Wessels, J. G. H. (1994) *EMBO J.*, **13**(24), 5848.
28. Lumsdon, S. O., Green, J., and Stieglitz, B. (2005) *Colloids Surf. B Biointerfaces*, **44**(4), 172.
29. Wösten, H. A. B., van Wetter, M. A., Lugones, L. G., van der Mei, H. C., Busscher, H. J., and Wessels, J. G. H. (1999) *Curr. Biol.*, **9**(2), 85.
30. Wosten, H. A. B., and de Vocht, M. L. (2000) *Biochim. Biophys. Acta — Rev. Biomembr.*, **1469**(2), 79.
31. Butko, P., Buford, J. P., Goodwin, J. S., Stroud, P. A., McCormick, C. L., and Cannon, G. C. (2001) *Biochem. Biophys. Res. Commun.*, **280**, 212.
32. Butko, P., Goodwin, J. S., Bufford, J. P., Stroud, P., McCormick, C. L., and Cannon, G. C. (2001), *Biophys. J.*, **80**, 404A.
33. Mackay, J. P., Matthews, J. M., Winefield, R. D., Mackay, L. G., Haverkamp, R. G., and Templeton, M. D. (2001) *Structure*, **9**(2), 83.

34. de Vocht, M. L., Scholtmeijer, K., van der Vegte, E. W., de Vries, O. M. H., Sonveaux, N., Wosten, H. A. B., Ruysschaert, J. M., Hadziioannou, G., Wessels, J. G. H., and Robillard, G. T. (1998) *Biophys. J.*, **74**(4), 2059.

35. Kwan, A. H., Winefield, R. D., Sunde, M., Matthews, J. M., Haverkamp, R. G., Templeton, M. D., and Mackay, J. P. (2006) *Proc. Natl. Acad. Sci. U. S. A.*, **103**(10), 3621.

36. Wang, X., Shi, F., Wosten, H. A., Hektor, H., Poolman, B., and Robillard, G. T. (2005) *Biophys. J.*, **88**(5), 3434.

37. Dague, E., Alsteens, D., Latge, J. P., and Dufrene, Y. F. (2008) *Biophys. J.*, **94**(2), 656.

38. Torkkeli, M., Serimaa, R., Ikkala, O., and Linder, M. (2002) *Biophys. J.*, **83**(4), 2240.

39. Ritva, S., Torkkeli, M., Paananen, A., Linder, M., Kisko, K., Knaapila, M., Ikkala, O., Vuorimaa, E., Lemmetyinen, H., and Seeck, O. (2003) *J. App. Crystallogr.*, **36,** 499.

40. Hakanpaa, J., Linder, M., Popov, A., Schmidt, A., and Rouvinen, J. (2006) *Acta Crystallogr. D Biol. Crystallogr.*, **62**(Pt 4), 356.

41. Hakanpaa, J., Paananen, A., Askolin, S., Nakari-Setala, T., Parkkinen, T., Penttila, M., Linder, M. B., and Rouvinen, J. (2004) *J. Biol. Chem.*, **279**, 534.

42. Hakanpaa, J., Parkkinen, T. A. A., Hakulinen, N., Linder, M., and Rouvinen, J. (2004) *Acta Crystallogr. D Biol. Crystallogr.*, **60,** 163.

43. Hakanpaa, J., Szilvay, G. R., Kaljunen, H., Maksimainen, M., Linder, M., Rouvinen, J., Popov, A., and Schmidt, A. (2006) *Protein Sci.*, **15**(9), 2129.

44. Booth, D. R., Sunde, M., Bellotti, V., Robinson, C. V., Hutchinson, W. L., Fraser, P. E., Hawkins, P. N., Dobson, C. M., Radford, S. E., Blake, C. C., and Pepys, M. B. (1997) *Nature*, **385**(6619), 787.

45. Askolin, S., Linder, M., Scholtmeijer, K., Tenkanen, M., Penttila, M., deVocht, M. L., and Wosten, H. A. B. (2006) *Biomacromolecules*, **7**(4), 1295.

46. de Vocht, M. L., Scholtmeijer, K., van der Vegte, E. W., de Vries, O. M., Sonveaux, N., Wosten, H. A., Ruysschaert, J. M., Hadziloannou, G., Wessels, J. G., and Robillard, G. T. (1998) *Biophys. J.*, **74**(4), 2059.

47. De Vocht, M. L., Reviakine, I., Ulrich, W. P., Bergsma-Schutter, W., Wosten, H. A. B., Vogel, H., Brisson, A., Wessels, J. G. H., and Robillard, G. T. (2002) *Protein Sci.*, **11**(5), 1199.

48. Wang, X., De Vocht, M. L., De Jonge, J., Poolman, B., and Robillard, G. T. (2002) *Protein Sci.*, **11**(5), 1172.

49. Yaguchi, M., Pusztai-Cary, M., Roy, C., Surewicz, W. K., Carey, P. R., Stevenson, K. J., Richards, W. C., and Takai, S. (1993) Amino acid sequence and spectroscopic studies of Dutch elm disease toxin, ceratoulmin, in *Dutch Elm Disease Research: Cellular and Molecular Approaches* (ed. Sticklen, M. B., and Sherald, J. L.), Springer-Verlag, New York, pp. 152–170.

50. Whiteford, J. R., and Spanu, P. D. (2002) *Mol. Plant Pathol.*, **3**(5), 391.

51. Tagu, D., Marmeisse, R., Baillet, Y., Riviere, S., Palin, B., Bernardini, F., Mereau, A., Gay, G., Balestrini, R., Bonfante, P., and Martin, F. (2002) *Eur. J. Histochem.*, **46**, 23.

52. Rillig, M. C. (2005) *Pedobiologia*, **49**, 395.

53. Teertstra, W. R., Deelstra, H. J., Vranes, M., Bohlmann, R., Kahmann, R., Kamper, J., and Wosten, H. A. (2006) *Microbiology*, **152**(12), 3607.

54. vanWetter, M. A., Schuren, F. H. J., Schuurs, T. A., and Wessels, J. G. H. (1996) *FEMS Microbiol. Lett.*, **140**(2), 265.

55. Paris, S., Debeaupuis, J. P., Crameri, R., Carey, M., Charles, F., Prevost, M. C., Schmitt, C., Philippe, B., and Latge, J. P. (2003) *App. Environ. Microbiol.*, **69**(3), 1581.

56. Thau, N., Monod, M., Crestani, B., Rolland, C., Tronchin, G., Latge, J. P., and Paris, S. (1994) *Infect. Immun.*, **62**(10), 4380.

57. Kershaw, M. J., Wakley, G., and Talbot, N. J. (1998) *EMBO J.*, **17**(14), 3838.

Chapter 4

Silkmoth Chorion: A Natural Protective Amyloid

Stavros J. Hamodrakas and Vassiliki A. Iconomidou
Department of Cell Biology and Biophysics, Faculty of Biology, University of Athens, Panepistimiopolis, Athens 157 01, Greece
shamodr@biol.uoa.gr, veconom@biol.uoa.gr

In this chapter, we review recent evidence on silkmoth chorion protein synthetic peptide-analogues and on silkmoth chorion itself, documenting the function of silkmoth chorion, the major component of the eggshell, a structure with extraordinary physiological and mechanical properties and a biological analogue of a cholesteric liquid crystal, as a natural protective amyloid.

4.1 Introduction

Amyloid fibrils are derived from soluble forms of proteins or peptides that convert, under certain conditions, into ordered fibrillar aggregates, called amyloid fibrils. Amyloidoses are an intriguing group of diseases including Alzheimer's, Parkinson's, Huntington's, type I diabetes, prion diseases, and many others, which may be caused by protein aggregation and ensued amyloid fibril formation.[1,2] However, in certain cases, living organisms take advantage of the

The Functional Fold: Amyloid Structures in Nature
Edited by Suzi Jarvis and Anika Mostaert

ISBN 978-981-4267-40-3 (Hardcover), 978-981-4267-62-5 (eBook)
www.panstanford.com

inherent ability of proteins and peptides to form amyloid structures and generate novel and diverse biological functions.[2–4] This was noted after our proposal that natural, protective amyloids do, indeed, exist.[5] Apparently, there is no similarity between amyloidogenic proteins or peptides in terms of amino acid sequence, molecular weight, morphology, or surrounding conditions.[2] In addition, many proteins, not implicated in amyloidoses, also called conformational diseases, have also been shown to form amyloids *in vitro*,[1,6] which led to the hypothesis that the potential for amyloidogenesis may be a near universal feature of proteins[6]. However, recent data indicate that amyloid formation is inherent in short stretches of amino acid sequences, called amyloidogenic determinants[7,8] and we have shown that it was derived, in certain cases, after millions of years of molecular evolution.[5,9]

The insect eggshell (chorion and vitelline membrane) and the associated follicular epithelium, which secretes its constituent layers, have been the subject of numerous investigations at the cellular and molecular level. These structures have provided a model system in several areas of current biological research, for example, for the physiology of the eggshell layers and morphogenesis of supramolecular structure, control of gene expression in differentiating cells, evolution of multigene families, and structural protein folding and organization.[10,12] The ovary of an adult female silkmoth usually consists of ovarioles, that is, strings of follicles attached to each other in linear arrays (Fig. 4.1a). The schematic diagram of each follicle is shown in Fig. 4.1b. The follicular cells, which surround the oocyte, synthesize, and secrete, according to a precise spatial and temporal program, a set of structural proteins onto the surface of the oocyte, which self-assemble to form the multilayered eggshell (Fig. 4.1b).[10,12]

The major component (90–95%) of the eggshell of many insect and fish eggs is chorion. Proteins account for more than 95% of its dry mass. This proteinaceous shell forms the outer layer of the eggshell and has extraordinary mechanical and physiological properties, protecting the oocyte and the developing embryo from a series of environmental hazards such as temperature variations, mechanical pressures, proteases, bacteria, and viruses.[10,12] It also allows for sperm entry and fertilization and for the exchange of the respiratory gases.[10,12] Figures 4.2 and 4.3 show electron micrographs of thin transverse sections of a silkmoth chorion. A lamellar ultrastructure of packed fibrils is seen: silkmoth chorion is a biological analogue of a cholesteric liquid crystal.[11,12]

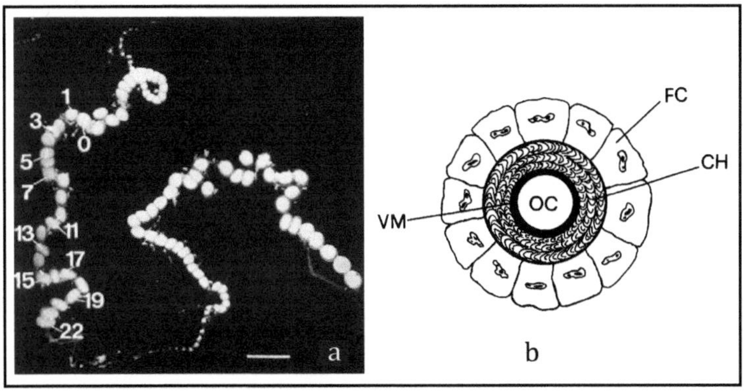

Figure 4.1 (a) Two ovarioles dissected from a developing adult female *Bombyx mori* moth. The paired ovaries in *B. mori* are each composed of four ovarioles. Follicles in progressively advanced stages of development (0, 1, 2, ..., 22) are interconnected in a linear array within each ovariole; 0 follicle indicates the beginning of choriogenesis (Papanicolaou *et al.* 1986). Bar = 3 mm. Copyright 2008 NRC Canada or its licensors. Reproduced with permission. (b) Schematic diagram of a follicle. FC, follicle cells; OC, oocyte. The eggshell consists of chorion (CH) and vitelline membrane (VM).

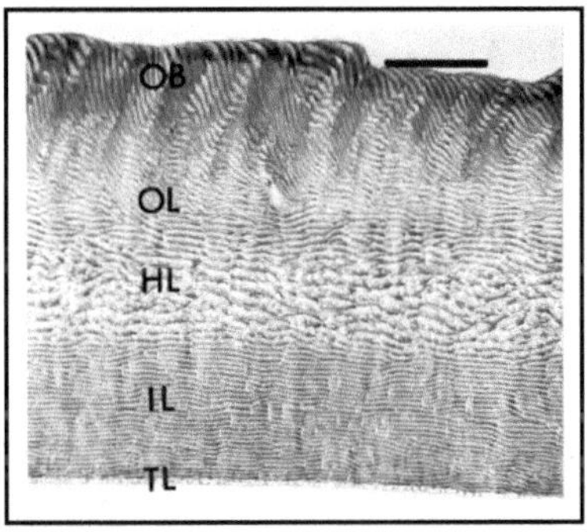

Figure 4.2 Transmission electron micrograph of a thin section cut through a chorion of *Abtheraea polyphemus* silkmoth. Within the bulk of the lamellar chorion, four types of lamellae can be distinguished: thin lamellae of the inner lamellar layer (IL); thick, distorted, spongy lamellae of the holey layer (HL); lamellae of the outer layer (OL); and, lying at an angle to the rest, the thick lamellae of the oblique layer (OB). Nearest to the oocyte is the trabecular (TL) layer, which consists of pillars surrounding air-filled spaces (Hamodrakas *et al.* 1986). Bar = 10 μm.

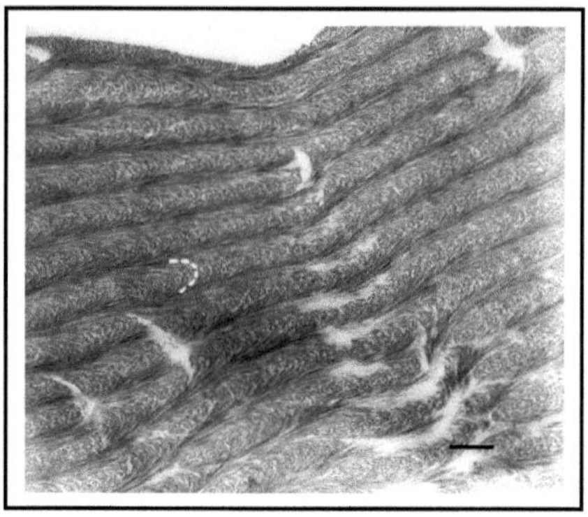

Figure 4.3 Transmission electron micrograph of a thin section cut through an *A. polyphemus* silk moth chorion, showing the fibrous ultrastructure of its lamellae. The parabolic pattern of fibres (approximately 110 Å; dotted lines) within each lamella, in oblique sections, is seen, which indicates that silkmoth chorion is a biological analogue of a cholesteric liquid crystal.[11,12] Bar = 0.3 μm.

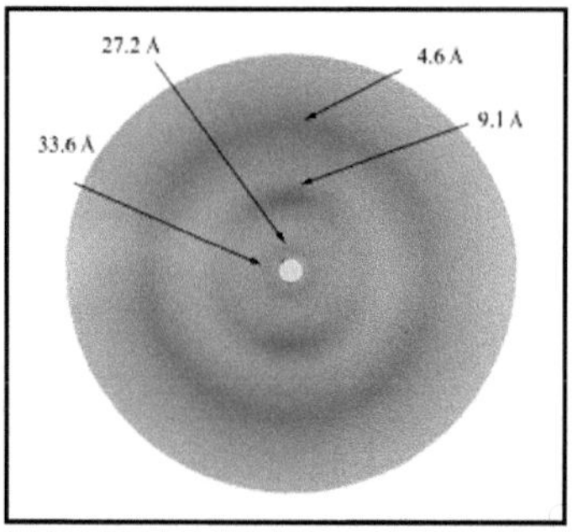

Figure 4.4 High angle X-ray diffraction pattern from an almost flat fragment of a mature silkmoth chorion of *A. polyphemus*. Incident beam is parallel to the chorion surface, which is horizontal. The plane of the X-ray film is vertical. Note the presence of 4.6, 9.1 and approximately 30 Å reflections, indicative of a β-sheet structure for silk moth chorion proteins. The 4.6 Å reflection corresponds to the repeat between hydrogen bonded molecular polypeptide chains of chorion proteins in a β-sheet conformation. The preferential orientation of the 9.1 Å reflection indicates a preferential orientation of stacked β-sheets parallel to the chorion surface, whereas the elliptical scattering at approximately 30 Å indicates a helicoidal architecture for silkmoth chorion and probably arises from approximately 30 Å protofilaments, constituents of the approximately 100 Å fibrils (Hamodrakas, 1992). A toroidal camera was employed.

The X-ray diffraction pattern of a silkmoth chorion shown in Fig. 4.4 indicates that β-sheet is the dominant secondary structure of its constituent proteins. FT-Raman spectroscopy (Fig. 4.5) and attenuated total reflectance Fourier transform infrared (ATR-FTIR) spectroscopy (Fig. 4.6) suggest that the β-sheets are antiparallel.[12]

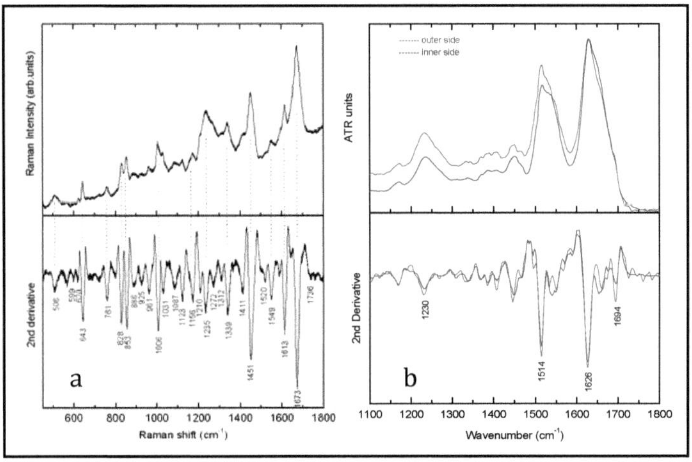

Figure 4.5 (a) FT-Raman (450–1800 cm^{-1}) spectrum of a silkmoth chorion of *A. polyphemus.* The second derivative spectrum is included. Error bar equals to 0.5σ. The positions of the amide I at 1673 cm^{-1} and amide III at 1235 cm^{-1} bands in the Raman spectrum are well known indicators of β-sheet secondary structure for silkmoth chorion proteins (Hamodrakas *et al.,* 1982). The β-sheets are very uniform and very well organised as can be judged by the rather narrow bandwidth (~30 cm^{-1}), at half maximum intensity, of the amide I band at 1673 cm^{-1}. (b) ATR-FTIR spectra taken from the outer (magenta) and inner (green) surfaces of an *A. polyphemus* silkmoth chorion. The second derivatives are included. The amide I and III bands at 1626 and 1230 cm^{-1}, respectively, clearly indicate a β-sheet type of structure for silk moth chorion proteins. The shoulder in the amide I region, at 1694 cm^{-1}, suggests that the β-sheets are anti-parallel. Reprinted from Biopolymers (Biospectroscopy), 72, 185–192, copyright (2003), with permission from John Wiley and Sons. See also Colour Insert.

About 200 proteins have been detected in the silkmoth chorion.[13] These proteins have been classified into two major classes, A and B.[10] The gene families encoding these proteins are related and constitute a superfamily with two branches, the α-branch and the β-branch.[14]

Sequence analyses and secondary structure prediction revealed that chorion proteins have a tripartite structure (Fig. 4.7), in that they consist of three domains.[12,15] The central domain is conserved in each of the two classes. The flanking N- and C-terminal domains are more variable and contain characteristic tandem repeats.[12,15] A and B central domains show distant similarities suggesting that the chorion genes constitute a superfamily derived from a single ancestral gene.[14]

The study of the properties of chorion proteins has long been hampered by the fact that it has proven very difficult to purify individual chorion proteins in large enough amounts of sufficient purity for structural studies. Therefore, several chorion protein peptide-analogues were synthesized and their structural and assembly properties were studied under various conditions.[5,9,16–19] These studies, which are partly reviewed in this work, suggest that silkmoth chorion is a natural protective amyloid.

4.2 Silkmoth Chorion Protein Peptide-Analogues

Initially, a 51-residue peptide was synthesized,[16] that can be considered as a generic central domain of the A class of silkmoth chorion proteins (Fig. 4.6). This peptide, referred to below as cA peptide, is representative for about 20–30% of all the proteinaceous

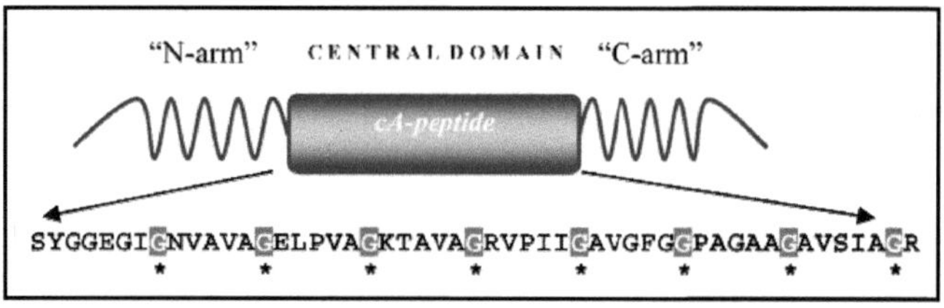

Figure 4.6 A schematic representation of the tripartite structure of silkmoth chorion proteins of the A family. A highly conservative central domain of invariant length, and two more variable flanking "arms" constitute each protein. Characteristic, tandemly repeating peptides are present both in the central domain and in the "arms" (Hamodrakas, 1992, and references therein). The amino acid sequence and relative position of the synthetic cA peptide (one letter code), designed to be an analogue of the entire central domain of the A family, is shown. Invariant glycines (G), repeating every six residues, are boxed and marked with an asterisk below the sequence.

material in the eggshell.[12] We chose this peptide because the central domains of the A class chorion proteins are highly conserved in both sequence and length and because this conservation indicates that this domain plays an important functional role in the formation of chorion structure.[5,12]

The cA peptide forms structurally uniform, amyloid-like fibrils by self-assembly in an astonishing variety of solvents, pH values, ionic strengths and temperatures (Iconomidou and Hamodrakas, in preparation). The fibrils were judged to be amyloid-like from their tinctorial and structural characteristics. They bind Congo red showing the characteristic for amyloids red-green birefringence when seen under crossed polars,[18] as well as Thioflavin-T (data not shown). Electron micrographs (Figs. 4.7 and 4.8) show that they are straight, unbranched double helices of indeterminate length and uniform in diameter (~100 Å). Each double-helical fibril consists of two protofilaments wound around each other. The protofilaments both have a uniform diameter of approximately 30–40 Å (Figs. 4.7 and 4.8).

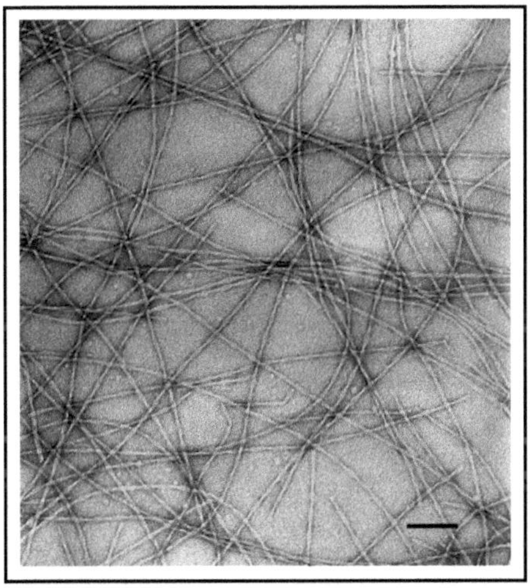

Figure 4.7 Electron micrograph of amyloid-like fibrils derived by self-assembly, from a 10 mg/ml solution of the cA peptide in a sodium acetate 50 mM buffer, pH 5. Fibrils are negatively stained with 1% uranyl acetate. They are of indeterminate length (several microns), unbranched, approximately 100 Å in diameter and have a double-helical structure. A pair of protofilaments, each 30–40 Å in diameter, are wound around each other, forming the double-helical fibrils. Bar = 100 nm.

Suspensions of these fibrils form oriented fibres, which give characteristic "cross-β" X-ray diffraction patterns[5] (Fig. 4.9). In these oriented fibres, the long axes of the amyloid-like fibrils seen in the electron micrographs of Figs. 4.7 and 4.8 are oriented more or less parallel to the fibre axis.

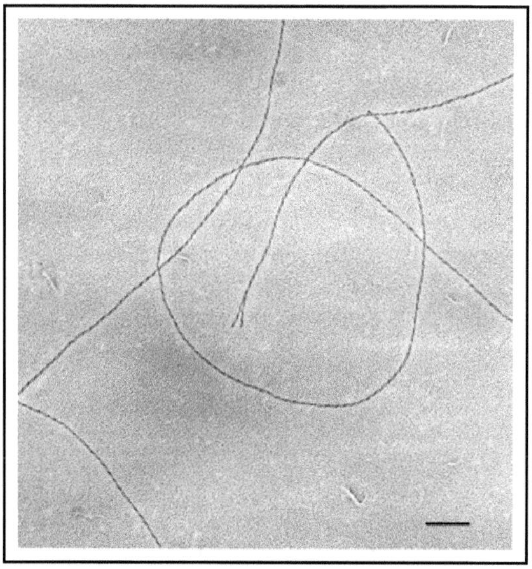

Figure 4.8 Electron micrograph of amyloid-like fibrils derived from a solution of the cA peptide (experimental conditions as in Fig. 4.7). Fibrils are rotary shadowed with Pt/Pd at an angle of 7 degrees under high vacuum. Bar = 1000 Å.

The oriented X-ray pattern (Fig. 4.9) taken from these fibres indicates the presence of oriented β-sheets in the amyloid-like fibrils of peptide cA. The presence of reflections corresponding to periodicities of 4.66 and 10.12 Å indicates the existence of β-sheets.[20] The strong meridional reflection at 4.66 Å suggests that the β-sheets are oriented so that their β-strands are perpendicular to the fibre axis and thus also to the long axis of the amyloid-like fibrils. The strong equatorial reflection at 10.12 Å, which corresponds to the intersheet distance, suggests that the packing of the β-sheets is parallel to the fibre axis and preferentially oriented. This X-ray pattern closely resembles typical cross-β patterns[21] taken from amyloid fibres (ref. 22, and references therein).

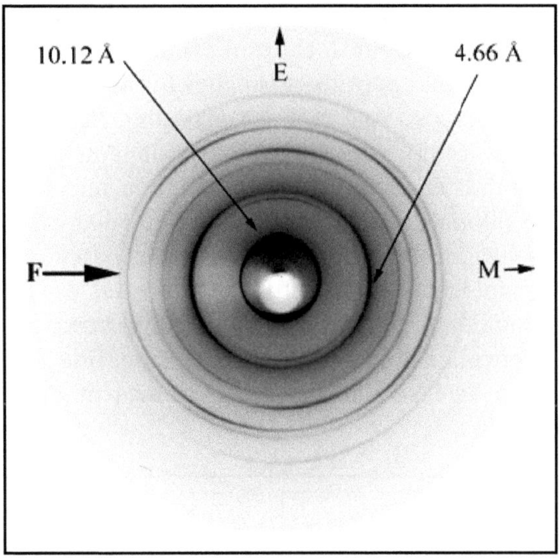

Figure 4.9 X-ray diffraction pattern from an oriented fibre of cA peptide amyloid-like fibrils. The meridian, M (direction parallel to the fibre axis) is horizontal and the equator, E, is vertical in this display. The X-ray diffraction pattern is a typical "cross-β" pattern showing a 4.66 Å reflection on the meridian and a 10.12 Å reflection on the equator. This indicates a regular structural repeat of 4.66 Å along the fibre axis (meridian) and a structural spacing of 10.12 Å perpendicular to the fibre axis. The structural repeat of 4.66 Å along the fibre axis corresponds to the spacing of adjacent β-strands (which should be perpendicular to the fibre axis, a "cross-β" structure), and the 10.12 Å spacing perpendicular to the fibre axis corresponds to the face-to-face separation (packing distance) of the β-sheets.

In addition, the FT-Raman and ATR FT-IR spectra of cA peptide amyloid fibrils, cast on a Au mirror (Fig. 4.10), clearly suggest that β-sheets are present in the structure of the cA peptide amyloid fibrils,[23] and, furthermore, that the β-sheets are antiparallel (shoulder at 1692 cm^{-1} in the amide I band).[18,24]

Further, we have designed and synthesized peptides with lengths equal to fractions of the length of the cA peptide. Thus, a 24-residue peptide was synthesized,[9] along with peptides of 12-residues and 6-residues in length (Iconomidou and Hamodrakas, in preparation). All these peptides fold and self-assemble, forming amyloid fibrils almost identical in structure and properties to the fibrils formed by the cA peptide. This clearly suggests that amyloid fibril formation propensity is inherent of the hexapeptide-repeating structure of the central domain of silk moth chorion proteins. This is further supported by the fact that peptide-analogues of the B family of silk moth chorion proteins also fold and self-assemble forming amyloid-like fibrils.[17] Thus, it appears that the amyloidogenic propensity in the case of silkmoth chorion proteins is hidden into the hexapeptide repeats of silkmoth chorion proteins, encoded after millions of years of molecular evolution.[5] Further, it probably implies that the underlying molecular substructure that dictates proper folding and self-assembly of chorion fibrils into the superstructure of silkmoth chorion is encoded into the tandem hexapeptide repeats present in the amino acid sequences of the central domain of silkmoth chorion proteins.

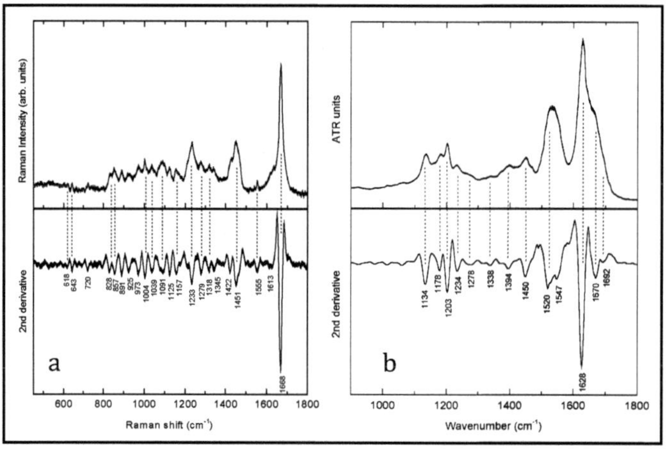

Figure 4.10 (a) FT-Raman (450–1800 cm^{-1}) spectrum of cA peptide amyloid fibrils, cast on an Au mirror. The second derivative spectrum is included. The positions of the amide I at 1668 cm^{-1} and amide III at 1233 cm^{-1} bands in the Raman spectrum suggest a β-sheet secondary structure for the cA peptide in the fibrils.[23] Error bar equals 0.5σ. (b) ATR FT-IR (900–1800 cm^{-1}) spectrum of cA peptide amyloid fibrils, cast on an Au mirror. The second derivative spectrum is included. The amide I and III bands at 1628 and 1234 cm^{-1}, respectively, clearly indicate a β-sheet type of structure for the cA peptide in the fibrils. The shoulder in the amide I region, at 1692 cm^{-1}, suggests that the β-sheets are antiparallel.[18,24] Error bar equals σ in the IR spectrum. Reprinted from Biopolymers (Biospectroscopy), 72, 185–192, copyright (2003), with permission from John Wiley and Sons.

At this point, it is relevant to mention how amyloid fibrillogenesis of silkmoth chorion protein peptide-analogues is actually performed. The phenomenon of the transformation of proteins into amyloid fibrils is of interest, not only because it is related to the protein folding problem, but also, because it is connected to the so-called conformational diseases, the amyloidoses. Consequently, various attempts have been directed towards an understanding of the fibrillogenesis pathway(s), with the aim of developing inhibitor drugs of therapeutic benefit. These are summarized in several excellent recent reviews.[6,25–29] However, the molecular and energetic factors affecting protein misfolding and amyloid fibrillogenesis are still largely unknown.[28,30]

Recently, we presented data, which clearly show that the first main step of amyloid-like fibrillogenesis from silk moth chorion peptides is the formation of nuclei of liquid crystalline nature.[18] Subsequently, these liquid-crystalline nuclei "collapse" and they are transformed into amyloid-like fibrils in a time period, that depends on several factors. The transformation is performed, most probably, as a result of a conformational transition to the structure of chorion peptides, from a left-handed parallel β-helix to an antiparallel β-pleated sheet (Iconomidou, Chryssikos, Gionis, and Hamodrakas, in preparation). Chorion peptides, apparently, have been designed to play this role after millions of years of molecular evolution.

4.3 Model Structure of Silkmoth Chorion Amyloidogenic Peptides

Taking into account all experimental and theoretical evidence accumulated previously for silkmoth chorion proteins[12] and their synthetic peptide analogues,[5,9,17] and the hexapeptide periodicities in the sequences of the A and B families of silkmoth chorion proteins,[31,32] the models shown in Figs. 4.11 and 4.12 for the peptide cA are the most probable models for the structure of silkmoth chorion peptides. The model shown in Fig. 4.11 is an antiparallel twisted β-pleated sheet of four-residue β-strands alternating with type II' β-turns. Invariant glyceine residues occupy the second position of the β-turns, a location especially favourable for glyceine in II' turns of twisted β-sheets of globular proteins.[33]

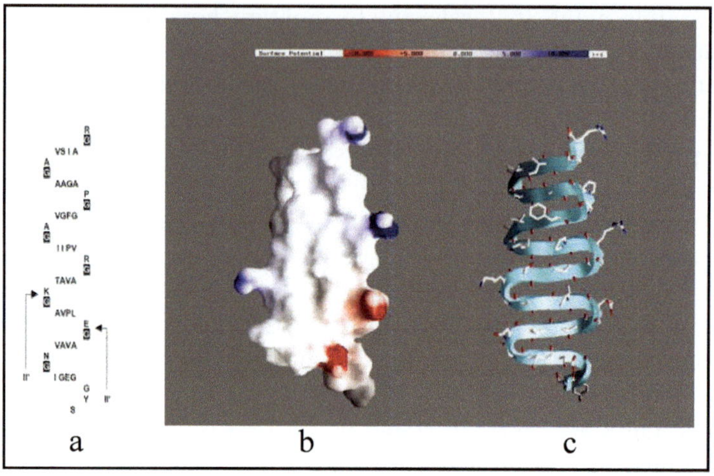

Figure 4.11 (a) Antiparallel, twisted β-sheet model proposed for the cA peptide. Sequence should be read continuously, beginning at the bottom. Invariant glycines (G) occupying the second position in the β-turns are black boxed. Tentative type II' β-turns alternate with four-residue β-strands. (b) A representation of the electrostatic surface potential of the antiparallel twisted β-sheet model proposed for the cA peptide in (a), calculated by the program DELPHI and displayed using GRASP (Nicholls *et al.*, 1991; Nicholls, 1993). The calculations were performed with the default charges file. Electrostatic potential is shown from −10 kT (red) to +10 kT (blue). (c) A ribbon model of the antiparallel twisted β-sheet model structure proposed for the cA peptide in (a), displayed using GRASP (Nicholls *et al.*, 1991). The side chains of the residues and the carbonyl oxygens of the main chain are shown as "sticks". View perpendicular to a "face" of the β-sheet. Reprinted from ref. 18, copyright (2004), with permission from Elsevier. See also Colour Insert.

Another possibility for the structure of the cA peptide might be that of the left-handed parallel β-helix (Fig. 4.12b), similar to that found in the structure of UDP-N-acetylglucosamine acyltransferase[34] and other left-handed parallel β-helical proteins.[4] This protein shows hexapeptide sequence motifs (Fig. 4.12a). It is interesting to note that right-handed, parallel β-helices, similar to those found in the pectate lyases, have been postulated as the main molecular components of amyloid protofibrils, although no detailed molecular models were presented.[35] Characteristic hexapeptide periodicities of both glycine and hydrophobic residues also appear in the sequence of the cA peptide.[12,31] Its sequence shows structural similarities with the sequence of UDP-N-acetylglucosamine acyltransferase (Fig. 4.12a), and the peptide would have a hydrophobic core when

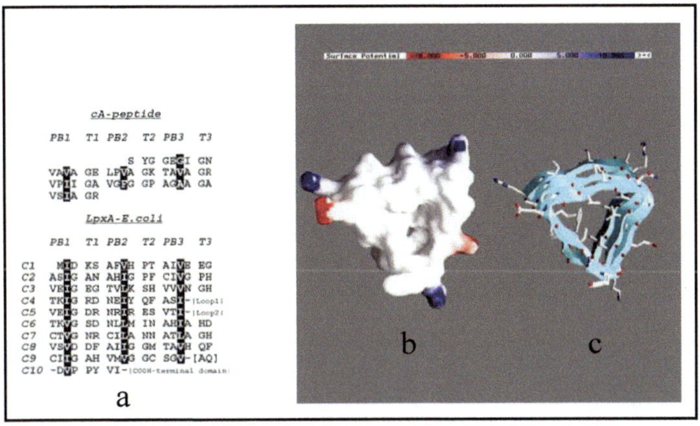

Figure 4.12 (a) Alignment of the sequence of the cA peptide with the sequence of the N-terminal domain of UDP-N-Acetylglucosamine Acyltransferase (LpxA of *Escherichia coli*) [Raetz, C. R. H. and Roderick, S. R. (1995) *Science* **270**, 997–1000]. The one letter code is used. Sequences should be read continuously beginning at the top. The structure of the N-terminal domain of LpxA of *E. coli* is a left-handed parallel β-helix [Raetz, C. R. H. and Roderick, S. R. (1995) *Science* **270**, 997–1000]. C1 to C10 rows identify individual turns of the helix. PB1, PB2, and PB3 denote the parallel β-strands of each turn. T1, T2, and T3 denote turn residues. Conserved hydrophobic residues, which have their side-chains pointing towards the interior of the left-handed parallel β-helix to form the hydrophobic core, are black boxed. (b) A representation of the electrostatic surface potential of the left-handed parallel β-helix model, derived by homology modelling for the cA peptide, from the alignment shown in (a) (Iconomidou *et al.*, 2000), calculated by the program DELPHI and displayed using GRASP (Nicholls *et al.*, 1991; Nicholls, 1993). The calculations were performed with the default charges file. Electrostatic potential is shown from −10 kT (red) to +10 kT (blue). (c) A ribbon model of the left-handed parallel β-helix model for the cA peptide, derived as described in (b), displayed using GRASP (Nicholls *et al.*, 1991). Tentative type II β-turns alternate with four-residue β-strands. The side chains of the residues and the carbonyl oxygens of the main chain are shown as "sticks". View almost parallel to the axis of the left-handed parallel β-helix. See also Colour Insert.

folded this way (Fig. 4.12b). Another attractive feature of the left-handed parallel β-helix model is the hydrophobic faces of the triangular prism-like β-helix (Fig. 4.12b). Nevertheless, the "edges" of this prism are occupied by charged, polar residues and glycines, and this makes 3-D packing difficult, unless there are very specific interactions. In contrast, the hydrophobic faces of the antiparallel β-sheet structure shown in Fig. 4.12 facilitate uniform 3-D packing of the β-sheets, leaving the polar and charged residues on both lateral "edges" of the sheet for favourable lateral interactions.

To our knowledge, we proposed the first structural model at atomic resolution for a left-handed parallel β-helix structure underlying amyloid fibrils.[5] The existence of β-helices in the amyloid fibrillar structure was later further described by several other groups.[4,36–41]

Clear support to the antiparallel twisted β-pleated sheet model shown in Fig. 4.11, as the prevalent structure present in amyloid fibrils formed from peptide-analogues of silkmoth chorion proteins is provided by the following: (i) preliminary calculations of X-ray diffraction patterns from the models presented in Figs. 4.11 and 4.12 and comparison with the experimental diffraction pattern of Fig. 4.8 (data not shown); (ii) analysis of the amide I band of FT-Raman spectra (Fig. 4.10; data not shown); (iii) the presence of the high frequency component in the 1690–1700 cm^{-1} in the ATR FT-IR spectra (Fig. 4.10) taken from samples containing amyloid-like fibrils formed from all, similar to natural, chorion peptides synthesized so far.

4.4 Silkmoth Chorion: A Natural Protective Amyloid

We have shown that cA-peptide fibrils have an amyloid nature, as well as all the other synthesized peptide-analogues of silkmoth chorion proteins, but we have so far silently assumed that the peptide fibrils are truly representative of the structure of chorion proteins in the eggshell. The cA peptide itself alone corresponds to approximately 25–30% of the total chorion mass. Its self-assembly mechanisms produce amyloid-like fibrils under a diverse variety of conditions, which strongly suggests that it should fold in an amyloid fashion in the physiological state, also *in vivo*. Concomitant evidence for this assumption can be found in Fig. 4.3. Lamellae (layers) of fibrils with the same dimensions (approximately 100 Å; see also refs. 10 and 11) as the cA peptide double-helical fibrils shown in Figs. 4.8 and 4.9 constitute the helicoidal architecture of silk moth chorion. Chorion fibrils consist of 30–40 Å protofilaments with a helical structure (ref. 12 and references therein). Furthermore, antiparallel β-pleated sheet is the dominant molecular conformation of silk moth chorion proteins *in vivo* (ref. 12 and references therein; Figs. 4.4–4.6). In addition, silkmoth chorion binds Congo red showing the characteristic red-green/yellow birefringence when seen under crossed polars (Fig. 4.13). This strengthens our proposal that silkmoth chorion is a natural protective amyloid.

Amyloids are generally associated with diseases such as Alzheimer's, spongiform encephalopathies, and type II diabetes etc. More than 20 types of human diseases are associated with the deposition of protein fibrils forming amyloid and resulting in tissue damage and degeneration.[1,2] Amyloidogenic proteins appear to be related by their ability to undergo a conformational change and adopt a new amyloidogenic conformation under partially denaturing conditions *in vivo*, which permits self-assembly into amyloid.[42,43] Our studies were the first to show that not all amyloids are, by definition harmful.[5] In chorion protein amyloids, the amyloidogenic conformation is, apparently, the native conformation.

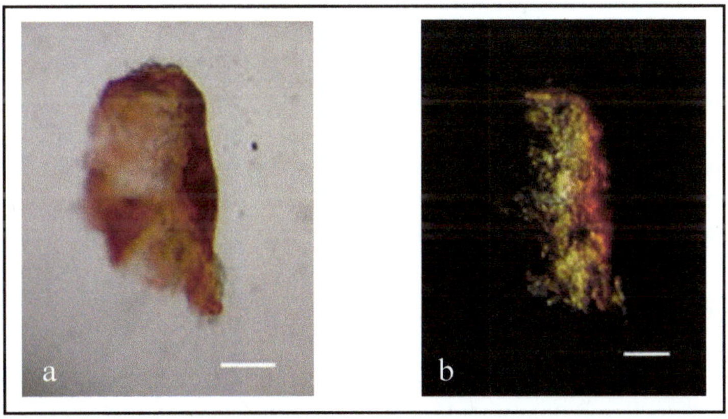

Figure 4.13 Photomicrographs of a part of a silkmoth chorion from *Bombyx mori* stained with Congo red: (a) bright field illumination, (b) crossed polars. The red-green/yellow birefringence characteristic for amyloids is clearly seen. Bar = 400 μm. See also Colour Insert.

Our proposal for the existence of natural, functional (in our case protective) amyloids[5] was followed by a number of examples, which undoubtedly confirmed the existence of several natural, functional amyloids, performing various important functions (e.g., refs. 2–4 and references therein, as highlighted in this book). We have recently reviewed the evidence for natural protective amyloids.[19]

To our knowledge, this is the first well-documented case where amyloid-like fibrils are formed from peptides that have a sequence so clearly folded in an antiparallel β-pleated sheet type of structure of the "cross-β" type. Nature, after millions of years of molecular evolution, has designed these amyloid-like chorion peptides to play an important functional role: to protect the oocyte and the developing embryo from a wide range of environmental hazards.[10,12] Chorion proteins self-assemble extracellularly to form the chorion of silk moths, a biological analogue of a cholesteric liquid crystal, far away from the follicle cells that synthesize and secrete them.[10–12] Chorion proteins and peptide-analogues provide a model system for the study of amyloid formation, with the potential to extract medically relevant information from the chorion destruction mechanisms[44] used by the embryo upon hatching.

Acknowledgements

We thank (in alphabetical order) Dr. B. Agianian, Dr. G. Chryssikos, Prof. P. Cordopatis, Mr. P. Everitt, Dr. A. Galanis, Dr. V. Gionis, Mr. K. Goldie, Prof. A. Hoenger, Dr. K. Leonard, Mr. C. Roome, Dr. L. Serrano, Dr. P. Tucker, and Prof. G. Vriend for their help and friendship, all these years. Special thanks to Prof. F.C. Kafatos for his unfailing interest and help. We also thank the University of Athens, the Greek Ministry of Research and Technology, the EMBL summer visitor's program and EMBO for financial support.

References

1. Uversky, V. N., and Fink, A. L. (2004) *Biochim. Biophys. Acta*, **1698**, 131.
2. Chiti, F., and Dobson, C. M. (2006) *Annu. Rev. Biochem.*, **75**, 333.
3. Kelly, J. W., and Balch, W. E. (2002) *J. Cell Biol.*, **161(3)**, 461.
4. Stevens, F. J. (2004) *J. Protein Folding Disor.*, **11**, 233.
5. Iconomidou, V. A., Vriend, G., and Hamodrakas, S. J. (2000) *FEBS Lett.*, **479**, 141.

6. Dobson, C. M. (1999) *Trends Biochem. Sci.*, **24**, 329.
7. Esteras-Chopo, A., Serrano, L., and de la Paz, M. L. (2005) *Proc. Natl. Acad. Sci. U.S.A.*, **102**(46), 16672.
8. de la Paz, M. L., and Serrano, L. (2004) *Proc. Natl. Acad. Sci. U.S.A.*, **101**(1), 87.
9. Iconomidou, V. A., Chryssikos, G. D., Gionis, V., Galanis, A. S., Cordopatis, P., Hoenger, A., and Hamodrakas, S. J. (2006) *J. Struct. Biol.*, **156**, 480.
10. Kafatos, F. C., Regier, J. C., Mazur, G. D., Nadel, M. R., Blau, H. M., Petri, W. H., Wyman, A. R., Gelinas, R. E., Moore, P. B., Paul, M., Efstratiadis, A., Vournakis, J. N., Goldsmith, M. R., Hunsley, J. R., Baker, B., Nardi, J., and Koehler, M. (1977) in *Results and Problems in Cell Differentiation*, vol. 8 (ed. Beerman, W.), Springer-Verlag, Berlin, Heidelberg, New York, p. 45.
11. Mazur, G. D., Regier, J. C., and Kafatos, F. C. (1982) in *Insect Ultrastructure*, vol. 1 (eds. Akai, H., and King, R. C.), Plenum Press, New York, p. 150.
12. Hamodrakas, S. J. (1992) in *Results and Problems in Cell Differentiation*, vol. 19 (ed. Case, S. T.), Springer-Verlag, Berlin, Heidelberg, p.115.
13. Regier, J. C., and Kafatos, F. C. (1985) in *Comprehensive Insect Biochemistry, Physiology and Pharmacology*, vol. 1 (eds. Gilbert, L. I., and Kerkut, G. A.) Pergamon Press, Oxford, New York, p. 113.
14. Lekanidou, R., Rodakis, G. C., Eickbush, T. H., and Kafatos, F. C. (1986) *Proc. Natl. Acad. Sci. U.S.A.*, **83**, 6514.
15. Hamodrakas, S. J., Jones, C. W., and Kafatos, F. C. (1982) *Biochim. Biophys. Acta*, **700**, 42.
16. Benaki, D. C., Aggeli, A., Chryssikos, G. D., Yiannopoulos, Y. D., Kamitsos, E. I., Brumley, E., Case, S. T., Boden, N., and Hamodrakas, S. J. (1998) *Int. J. Biol. Macromol.*, **23**, 49.
17. Iconomidou, V. A., Chryssikos, G. D., Gionis, V., Vriend, G., Hoenger, A., and Hamodrakas, S. J. (2001) *FEBS Lett.*, **499**, 268.
18. Hamodrakas, S. J., Hoenger, A., and Iconomidou, V. A. (2004) *J. Struct. Biol.*, **145**, 226.
19. Iconomidou, V. A., and Hamodrakas, S. J. (2008) *Curr. Prot. Pept. Sci.*, **9(3)**, 291.
20. Fraser, R. D. B., and MacRae T. P. (1973) *Conformation in fibrous proteins and related synthetic polypeptides*, Academic Press, New York, London.
21. Geddes, A. J., Parker, K. D., Atkins, E. D. T., and Beighton, E. (1968) *J. Mol. Biol.*, **32**, 343.

22. Sunde, M., and Blake, C. (1997) *Adv. Prot. Chem.*, **50**, 123.
23. Hamodrakas, S. J., Asher, S. A., Mazur, G. D., Regier, J. C., and Kafatos, F. C. (1982) *Biochim. Biophys. Acta*, **703**, 216.
24. Jackson, M., and Mantsch, H. H. (1995) *Crit. Rev. Biochem. Mol. Biol.*, **30**(2), 95.
25. Rochet, J. C., and Lansbury, Jr. P. T. (2000) *Curr. Opin. Struct. Biol.*, **10**, 60.
26. Walsh, D. M., Hartley, D. M., Kusumoto, Y., Fezoui, Y., Condron, M. M., Lomakin, A., Benedek, G. B., Selkoe, D. J., and Teplow, D. B. (1999) *J. Biol. Chem.*, **274**, 25945.
27. Kelly, J. W. (2000) *Nat. Struct. Biol.*, **7**, 824.
28. Soto, C. (2001) *FEBS Lett.*, **498**, 204.
29. Žerovnik, E. (2002) *Eur. J. Biochem.*, **269**, 3362.
30. Carel, R. W., and Gooptu, B. (1998) *Curr. Opin. Struct. Biol.*, **8**, 799.
31. Hamodrakas, S. J., Etmektzoglou, T., and Kafatos, F. C. (1985) *J. Mol. Biol.*, **186**, 583.
32. Hamodrakas, S. J., Bosshard, H. E., and Carlson, C. N. (1988) *Prot. Eng.*, **2**(3), 201.
33. Sibanda, B. L., and Thornton, J. M. (1985) *Nature*, **316**, 170.
34. Raetz, C. R. H., and Roderick, S. R. (1995) *Science*, **270**, 997.
35. Lazo, N. D., and Downing, D. T. (1998) *Biochemistry*, **37**(7), 1731.
36. Wille, H., Mitchelitsch, M. D., Guénebaut, V., Supattapone, S., Serban, A., Cohen, F. E., Agard, D. A., and Prusiner, S. B. (2002) *Proc. Natl. Acad. Sci. U.S.A.*, **99**(6), 3563.
37. Wetzel, R. (2002) *Structure*, **10**, 1031.
38. Pickersgill, R. W. (2003) *Structure*, **11**, 137.
39. Williams, A. D., Portelius, E., Kheterpal, I., Jun-Tao Guo, Cook, K. D., Hu, Y., and Wetzel, R. (2004) *J. Mol. Biol.*, **335**, 833.
40. Kishimoto, A., Hasegawa, K., Suzuki, H. Taguchi, H., Namba, K., and Yoshida, M. (2004) *Biochem. Biophys. Res. Comm.*, **315**, 739.
41. Govaerts, C., Wille, H., Prusiner, S. B., and Cohen, F. E. (2004) *Proc. Natl. Acad. Sci. U.S.A.*, **101**(22), 8342.
42. Kelly, J. F. (1996) *Curr. Opin. Struct. Biol.*, **6**, 11.
43. Kelly, J. F. (1998) *Curr. Opin. Struct. Biol.*, **8**, 101.
44. Hagenmaier, H. E. (1974) *Comp. Biochem. Physiol.*, **49b**, 313.

Chapter 5

Spider Silk

Ute Slotta,* Kristina Spieß,* and Thomas Scheibel
Fakultät für Angewandte Naturwissenschaften, Lehrstuhl für Biomaterialien, Universität Bayreuth, 95440 Bayreuth, Germany
Thomas.scheibel@uni-bayreuth.de

Silks are a class of structural proteins produced by arthropods, which like amyloids are evolutionarily very ancient structures. Silk proteins are produced in special glands and are stored at high concentrations in solution known as dope until they are transformed into a solid silk thread. The processing conditions have a great influence on the protein structure and the mechanical properties of the final thread. Ancient cocoon silks (used for the preservation of offspring) display cross-β structures similar to amyloid fibrils. Spiders have evolved other silks for task-specific applications (such as the capture of prey or as a lifeline to escape from predators) with different mechanical properties. The spinning process evolved in parallel, becoming more complex. In contrast to ancient silks, more recent silks tend to display predominantly parallel-β structures, but under distinct processing conditions adopt amyloid-like structures.

*Kristina Spieß and Ute Slotta contributed equally to this chapter.

The Functional Fold: Amyloid Structures in Nature
Edited by Suzi Jarvis and Anika Mostaert

ISBN 978-981-4267-40-3 (Hardcover), 978-981-4267-62-5 (eBook)
www.panstanford.com

5.1 Introduction

Functional amyloids are typically secreted or extruded structural proteins which exist in various animal kingdoms like bacteria, fungi, invertebrates, and vertebrates. Besides its presence in functional amyloids, it is proposed that the amyloid structure represents a generic conformational state which can under certain circumstances be adopted by any polypeptide chain.[1,2]

In this article, we focus on typically non-amyloidogenic proteins, namely silk, a class of structural proteins solely produced by arthropods. Many different types of silks are known, each displaying distinct mechanical properties which are perfectly adapted to their application ranging from protection to reproduction purposes and capturing prey. Most arthropods can produce only one type of silk, whereas orb weaving spiders are able to produce up to 6 different kinds of silk and a glue substance (Fig. 5.1).

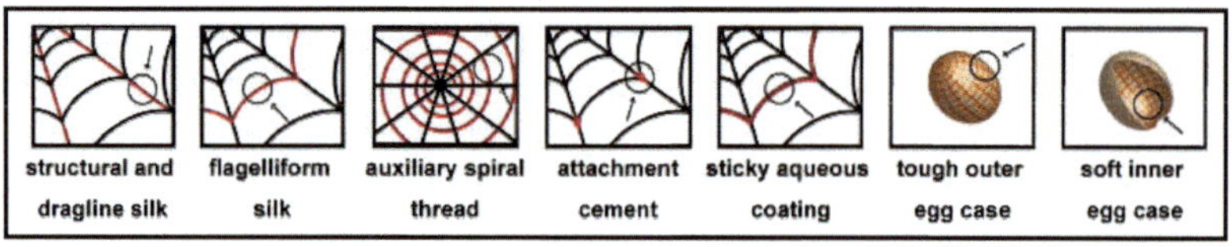

Figure 5.1 Different silks produced by a female orb weaving spider. Each is adapted perfectly to its purpose mainly regarding mechanical properties. The two major and best investigated silks are dragline silk and flagelliform silk. The dragline is a tough fibre forming frame and radii of the web, additionally serving as lifeline when the spider drops. Flagelliform silk is more elastic to dissipate the impact energy of flying prey. Adapted and reproduced from Romer and Scheibel,[3] with permission from Landes Bioscience. See also Colour Insert.

5.2 Overview of Silk Structure, Mechanics, and Assembly

The basic components of silks are specific proteins, typically stored in distinct glands which have evolved from one single type of gland.[4] Various silks reveal different structures as summarized in Table 5.1 (modified from Craig[5]).

Table 5.1 Selection of silk structures produced by insects and spiders

Protein structure	Protein characteristics	Silk origin
α-helical	Low glyceine content	Insects
	High content of acidic residues	Spiders (when ejected or extruded instead of spinning)
	Converts to parallel-β	
Parallel-β	Classified in 6 Warwicker groups, characterized by increasing side chain length and decreasing glyceine content	Insects
	Polypeptide chain oriented parallel to fiber axis	Spiders
Cross-β	High serine content	Insects
	Polypeptide chain oriented perpendicular to fiber axis	Spiders (*in vitro*, formed by a recombinant spider silk protein)[1,2]
	Converts to parallel-β structure under stress	

Most spider silks belong to the class of parallel-β silks, but upon ejection or extrusion of spider silk proteins also α-helical structures can be adopted. It has to be noted that additional structural subclasses exist, which are not discussed in context of this review.

The α-helical silks are mainly produced by insects like bees and fleas that do not possess glands that are entirely dedicated to silk production. Such silks are used for reinforcement, nets, pupation cases, or cocoons. Although no spider is known to produce α-helical silks, it is assumed that spitting spiders eject or spit silk with α-helical structure, for example. It has been found that α-helical silk can be converted into parallel-β structure under shear stress.[5,6]

Silks with parallel-β-pleated sheet structure showing either parallel or antiparallel orientation of the strands within the sheets are produced by both insects and spiders. Warwicker classified parallel-β-silks into six sub groups differing in length of the side chains and in their glycine content.[7]

A third type of silk structure is a cross-β-pleated sheet which is produced by insects like the lacewing *Chrysopa flava*. Cross-β silk is used by lacewing larvae for cocoon production as well by adults for attaching eggs to the lower side of leaves. This type of structure resembles that of amyloid fibrils but it can be converted into parallel-β silk upon stretching.[5,6]

There are three more structural silk classes which can be found among insects such as collagen-like silk, polyglycine II-like structures, and cuticulin-silk.[5]

Although all silks display good mechanical properties, spider dragline silks are superior in combining strength and elasticity, resulting in an extraordinary toughness which can match with that of high-performance synthetic polymers such as nylon or Kevlar (Table 5.2).

Table 5.2 The mechanical properties of different natural and synthetic high-performance fibres (numbers taken from Gosline *et al.*[8])

Material	Stiffness E_{init} (GPa)	Strength σ_{max} (GPa)	Extensibility ε_{max}	Toughness MJ m^{-3}
Dragline silk *(Araneus diadematus)*	10	1.1	0.27	160
Cocoon silk *(Bombyx mori)*	7	0.6	0.18	70
Collagen (Mammalian tendon)	1.5	0.15	0.12	7.5
Nylon fiber	5	0.95	0.188	80
Kevlar 49 fiber	130	3.6	0.027	50
Carbon fiber	300	4	0.013	25
High-tensile steel	200	1.5	0.008	6

Dragline silk combines high tensile strength (1.1 GPa) with good extensibility (0.27) yielding an outstanding toughness (160 MJ/m^3) succeeding that of all other displayed materials. Dragline silks are composed of proteins[9–12] which are produced in the major ampullate gland consisting of a long tail and a wide sac reflecting the compartment where the proteins are stored. Within the glands extremely high protein concentrations of up to 50% (w/v) exist forming the so-called spinning dope (Fig. 5.2).[13]

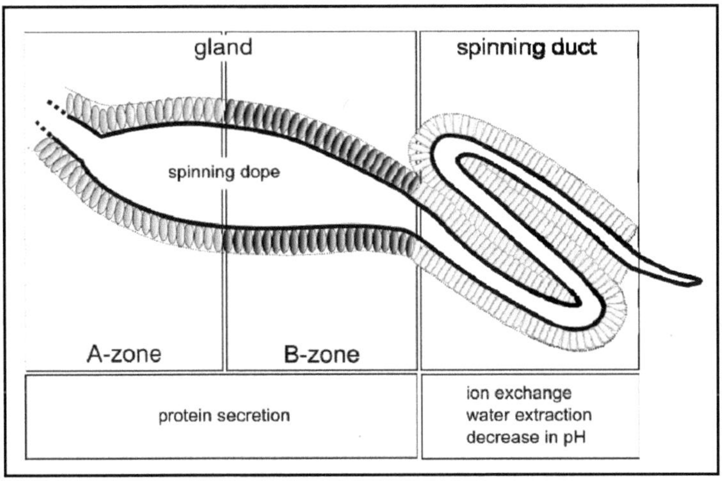

Figure 5.2 A spider's spinning gland is divided in an A-zone and a B-zone, where specialized cells secrete silk proteins. Structure formation and protein assembly are initiated in the spinning duct, where ion exchange, water extraction, pH decrease and shear forces are applied (reviewed in Hardy *et al.*[14] and Kenney *et al.*[15]).

Originating from the gland, the spinning dope enters the spinning duct consisting of three siphon shaped loops. Structure formation of the proteins already starts in the B-zone of the silk gland, where fibrillar structures have been detected reflecting amyloidogenic properties such as cross-β structure.[16] However, the presence of amyloid-like nanofibrils in the B-zone has to be treated with caution. Because the final dragline silk belongs to the structural class of parallel-β silks, the question arises how the structural conversion takes place and whether these fibrils play a role in thread formation or whether they reflect artefacts of sample preparation.

Within the spinning duct phase separation is induced by kosmotropic ions leading to removal of water and concentrating the proteinacous phase. Within the protein phase further folding takes place, additionally triggered by acidification and shear forces.[17–20]

Several methods used to investigate structure and assembly of fibrous proteins and to analyse amyloidogenic properties of protein fibrils can be employed for silk analysis. Studies by atomic force microscopy or transmission electron microscopy revealed that silk fibrils as detected in the B-zone or in *in vitro* assembly reactions, similar to amyloid fibrils, are straight, rigid, unbranched, and up to several micrometers long.[21]

Like amyloids, spider silk nanofibrils bind the dyes Congo red and thioflavin T, further indicating that silk fibrils show certain amyloidogenic characteristics.[21] Not only silk nanofibrils bind Congo red, but spider dragline silk derived from *Nephila edulis* was also found to bind this dye.[22] Since the specificity of dye binding depends on solution and staining conditions[23,24] (there are some examples of cross-reactions of Congo red with non-amyloidogenic aggregates, globular proteins, and even non-proteinaceous materials), Congo red binding has to be considered with caution.[25–27] A more detailed analysis of secondary structural features, especially of antiparallel or parallel β-sheets, can be obtained by Fourier transformed infrared (FTIR) spectroscopy, and the orientation of the strands forming β-sheets (cross-β) is typically monitored by X-ray fibre diffraction studies (see the following text).

5.2.1 Specific Structural Features of Silks

As mentioned earlier, silks of insects and spiders are classified according to their predominant secondary structure.[5,28] α-silks show a high content of α-helical structure, with a X-ray fibre diffraction (XRD) pattern similar to that obtained for α-keratin. Their amino acid composition differs significantly from that of other silks with a much lower content of glycine and an elaborated amount of charged residues.[5,28,29]

β-Silks are characterized by their high proportion of β-sheet structure as detected by methods like FTIR or Raman micro-spectroscopy. As an example, *Bombyx mori* cocoon silk contains 50% and *Nephila* dragline silk 37 % of β-sheet structure.[30] Analysing the maxima of the amide I band of FTIR spectra of dragline spider silk revealed peaks at 1637 cm^{-1},[31] of silk produced by the green lacewing *C. flava* at 1625 cm^{-1},[29] and of nanofibrils formed by recombinant spider silk at 1623 cm^{-1},[21] all of which are representative for β-sheet rich structures. Interestingly, the later wave numbers are within a range specific for amyloid fibrils.[21,32]

The arrangements of β-pleated sheets in β-silks can also be determined by XRD. XRD spectra of *C. flava* egg stalk silk reveal a series of layer lines corresponding to a (pseudo)repeat of 9.48 Å on the meridian and a strong equatorial reflection signal at 5.45 Å

(Fig. 5.3a), which can be assigned to an arrangement, where the individual β-strands of a β-sheet run perpendicular to the fibre axis, whereas the sheets are oriented in parallel to the axis.[29,33] The meridional reflection signal thereby represents the main-chain spacing reflecting the distance between two β-strands (where a distance of about 9.6 Å was reported to reflect two antiparallel chains with a marked repeat at the half distance at 4.8 Å), while the equatorial signal arises from the spacing between two adjacent β-sheets.[34,35] Geddes *et al.*[33] proposed a model where the β-strands are stacked with turns every eighth amino acid, resulting in a ribbon-like shape with 25 Å in diameter.

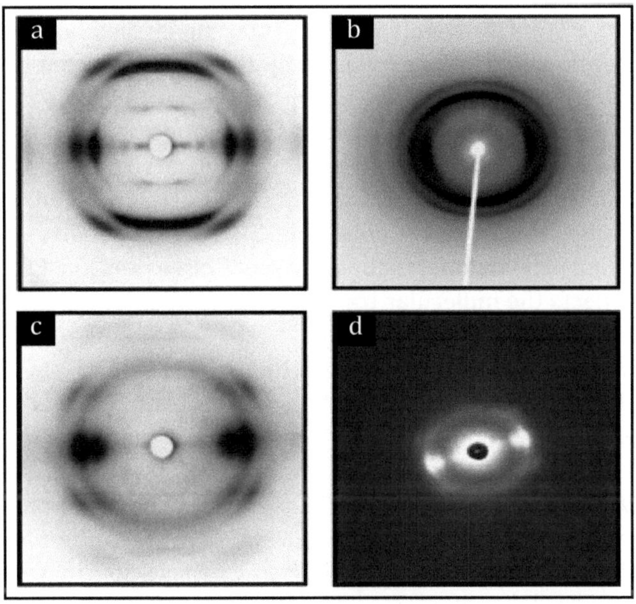

Figure 5.3 X-ray diffraction pattern of different silks. (a) Pattern of egg stalk silk of *Chrysopa flava* unstretched, displaying cross-β structure.[33] (b) Pattern of nanofibrils of a recombinantly produced spider silk protein displaying cross-β structure[21] (figure taken from Slotta *et al.*,[21] copyright Wiley-VCH Verlag GmbH & Co. KGaA, reproduced with permission). (c) Pattern of egg stalk silk of *C. flava* stretched, displaying parallel-β structure.[33] (a and c: reproduced with permission from Elsevier.) (d) Pattern of *Nephila clavipes* spider silk showing typical reflection signals for parallel-β structure (figure taken from: Parkhe *et al.*,[36] copyright John Wiley & Sons Limited, reproduced with permission).

The diffraction pattern described earlier is consistent with that of various denatured globular and fibrous proteins, and has been found to be present in every analysed amyloid structure.[28,35] Therefore, it is considered to be a common feature of the amyloid core structure. Nanofibrils formed from recombinant spider silk display a similar pattern (Fig. 5.3b), indicating the presence of a cross-β structure. A sharp meridional reflection at 4.7 Å is observed, as well as an equatorial spacing of 5.3 Å.[21] A specific feature of the silk diffraction patterns is the quite narrow inter-sheet distance (when compared to typical amyloid patterns), which is thought to result from less bulky amino acid side chains (there is a high content of glycine and alanine residues in silk proteins) in the β-strands forming the cross-β structure.[5,21,28] The slightly higher inter-sheet distance observed for *Chrysopa* silk can be explained by the higher content of serine residues.[5]

Interestingly, upon stretching of the *Chrysopa* silk to its six-fold length, a change in the diffraction pattern is observed.[33] The stretched silk gives rise to reflection signals in the equatorial direction, whereas the meridional signals are gone (Fig. 5.3c), with the degree of the effect being proportional to the degree of fibre extension. Likely, this effect reflects the molecular transformation of the cross-β structure to a parallel-β structure.[33]

The majority of silks produced by insects and spiders represents parallel-β silks with β-strands of individual sheets oriented parallel to the direction of the fibre axis.[5,28] As a result, XRD patterns consist only of equatorial reflections arising from both the main chain-and inter-sheet spacing (Fig. 5.3d). Depending on the amino acid residues involved in β-sheet formation and on the parallel or antiparallel orientation of the β-strands, these spacings vary for different parallel-β silks.[7] *B. mori* silk shows an inter-sheet spacing of 9.3 Å, assigned to a composition of alternating Ala-Gly residues.[5,37] Further analysis revealed that the methyl groups of the alanine residues alternately point to opposite sides of the sheet, resulting in an antipolar-antiparallel arrangement of the strands within a β-sheet.[30,38–40] Silks with β-sheets consisting of mostly alanine or serine residues consequently have bigger inter-sheet spacings ranging from 10.6 Å up to 15 Å.[5,37]

Solid state nuclear magnetic resonance (NMR) measurements and Raman spectroscopy verified these structural features by analysing polypeptides consisting of the amino acid motif GAGAGS of *B. mori* cocoon silk[30,41,42] and of polyalanine segments of spider draglines.[36,43,44] Recent studies revealed that beneath polyalanine stretches also GGA motifs are likely to be incorporated into β-sheet structures of dragline silk,[30] even though glycine-rich motifs such as GGA or GPGQQ preferentially do not form crystalline β-sheet structures.[43] Further, 3_1-helical structures (e.g., polyglycine II-helices), β-turns (type I and III), and some unordered structures were found in dragline silk, which are thought to build an amorphous matrix for the crystalline-like β-sheet structures.[15,30,45,46] Typical features of the amorphous matrix are a high content of glycine and, in *B. mori*, tyrosine residues, arranged in amino acid motifs like GAGAGY.[47] However, NMR studies indicated the existence of regions with different degree of orientation in the amorphous matrix, including less ordered/isolated β-sheets as well as aligned and pre-stressed chains.[43,44,48] Recent FTIR studies suggest a model where the highly oriented β-sheet crystals are interconnected by amorphous chains displaying a Gaussian distribution of pre-strain.[48–50]

This hierarchical composition also bears the explanation for the observed outstanding mechanical properties of spider silks. In general, the β-sheet crystallites are thought to account for the mechanical strength of the final thread and to be incorporated like rigid fillers in the amorphous region,[51] which most likely accounts for the elasticity. Indeed, silks with higher elasticity have been shown to contain a larger fraction of flexible structures like β-spirals or 3_1-helices.[30,52] Further, the degree of pre-strain of the interconnecting chains has been proposed to play a key role in controlling and fine-tuning the final characteristics.[53] Another parameter influencing the mechanical stability is thought to be the orientation of the crystallites, which can be triggered by the mode of spinning, with increasing spinning speed leading to a higher orientation within the thread.[54,55] During supercontraction of the fibre — a phenomenon which is observed upon wetting leading to shrinkage of the fibre accompanied by an increase in diameter — the thread is plasticized and some orientation is lost.[36,56] Supercontraction is followed by a more rubber-like behaviour of the fibre. However, after re-stretching, the crystals are thought to be re-aligned, resulting in the original degree of orientation.

5.2.2 Assembly of Silk Proteins

Very little is known about the assembly of cross-β silks. They are produced by some insect larvae in Malpighian tubules and, additionally, in colleterial glands of adult green lacewings (*C. flava*).[5,57] The colleterial glands are sac shaped and composed of columnar secretory cells containing an apical end apparatus reflecting infoldings of the plasma membrane. Analysis of sectioned material revealed the presence of sheet-like micelles which are several hundred angstrom wide and 25 Å thick, matching the dimensions of the ribbon-like structure calculated from XRD.[57] Malpighian tubules are a type of excretory and osmoregulatory system and represent an example for glands not primarily dedicated to silk production. As the system is usually applied to absorb water and contains active ion transporters, one might speculate that the later are also involved in silk assembly.

Parallel-β silks are much more thoroughly analysed and several aspects of their assembly process have been unravelled. Two of the best characterized parallel-β silks are the cocoon silk from *B. mori* and spider dragline silks from *Nephila* and *Araneus* species. The underlying proteins are produced in salivary glands in the case of the silkworm *B. mori* and the major ampullate glands in case of the spiders.[13] The assembly processes of both silk types are similar regarding the involved factors, therefore, for simplification, we will only focus on the assembly of spider dragline silk in more detail.

5.2.2.1 Crucial factors triggering spider silk protein assembly

After secretion from specialized, tall columnar endothelial cells into the lumen of the ampulla, dragline silk proteins, also called major ampullate spidroins, are stored mainly intrinsically unstructured or at least only partially structured at very high concentrations (up to 50% w/v) in the liquid dope (Fig. 5.2).[4] To form the final thread, they have to undergo a liquid — solid phase transition which is accompanied by the formation of β-sheet structures (as described earlier) stabilized by intra-and inter-chain hydrogen bonds.[58] Factors involved in the extension, alignment, and dense packing of the molecules include physical forces like increasing elongational flow in combination with biochemical changes in the spinning duct.[46,59] These are in detail an increase in the concentration of kosmotropic phosphate ions, while chloride ions are assumed to be extracted. The pH changes from 6.9 to 6.3, and the concentration of the proteins in the spinning dope increases by the extraction of water.[4,60]

To clarify the individual influence of these factors, the assembly of recombinant spider silk proteins was investigated using a microfluidic device.[61] Upon exposition to potassium phosphate, the proteins formed spherical aggregates.[61,62] The aggregates could be merged to finally form a fibre by simultaneously dropping the pH value and employing an elongational flow.[62] These precisely controlled triggers occurring in the distal part of the spinning duct are thought to assist in the assembly and locking of the final state of the protein molecules.[17–20] Thread formation *in vivo* is completed by stretching accomplished by pulling the silk. Recently, a salt-, pH-, and shear-sensitive dimerization behaviour of the short non-repetitive domains at the protein's termini has been demonstrated, stressing their relevance for fibre assembly.[17–20]

Finally, the pulling step reflects a fibre finishing procedure that depends on several parameters such as reeling speed, temperature, and relative humidity, which critically influence mechanical properties of the final thread.[15,22,54] The fact, that silks with various secondary structures, such as α-silks or cross-β silks, can be transformed into parallel-β silks upon mechanical stress[5,33] underlines that the mode of processing has a great influence on silk structure and its properties.

5.2.2.2 Models for silk assembly

It is assumed, that the amphiphilic nature of spider silk proteins leads to a pre-orientation of the proteins inside the ampulla, thereby supporting solubility and preventing premature aggregation. Three mechanisms concerning this pre-orientation or pre-alignment have been discussed in the literature.

The first mechanism involves the arrangement of the proteins into micelle-like structures. Small droplets were found in solutions collected directly from the spinning gland of *B. mori*.[63] *In vitro* investigations on regenerated silk fibroin originating from *B. mori*[63] and recombinantly produced spider silk proteins[17,19,61,64] support the theory of micelle formation. It is suggested that the hydrophilic non-repetitive regions of the amphiphilic proteins form the outer shell of such structures, while the more hydrophobic parts are buried inside. With increasing protein concentration, the observed micelles in the gland of *B. mori* fuse via coalescence, forming observable globular structures of 1–15 μm in diameter.[63] The presence of shear force, resulting from the elongational flow within the spinning duct, further leads to a stretching and elongation of the globules. An interaction of these "elongated globules" is proposed to be the basis for fibril formation upon further alignment of the molecules.[63]

The theory presented by Vollrath *et al.*[4] is based on the consideration of the spinning dope as a crystalline liquid, which is able to flow while displaying a high orientational order. Typical arrangements of biopolymers at sufficiently high concentration include nematic and hexagonal columnar phases.[4,65,66] In the major ampulla gland, the long axes of the spider silk proteins have been described to be oriented perpendicular to the epithelial wall right after secretion.[4] Approaching the centre of the ampulla the long axes of the molecules are aligned with the gland's wall (Fig. 5.4b). This arrangement, called an escaped nematic texture, is proposed to change again upon narrowing of the duct. Bending of the molecules backwards and forwards would lead to the formation of a bi-layered disc. Further application of shear forces and an accelerating elongational flow finally leads to an alignment of these structures. Upon changing the chemical parameters and water content, as described earlier, thread formation is completed.

In a recent review it was suggested that these two theories might not exclude each other, but rather reflect different systems investigated involving different protein concentrations.[14] While amphiphilic molecules will spontaneously form micelles at lower protein concentrations, they will arrange into hexagonal columns at higher concentrations, thereby representing a general characteristic behaviour of lyotropic liquid crystal systems involving amphiphilic molecules.[14,67]

A third theory is based on a nucleation-aggregation process similar to amyloid fibril formation.[68] It implies the formation of β-sheet aggregates as a result of the stress-induced conformational transition of some molecules into a β-sheet-rich structure followed by further self-assembly. These assemblies then act as nuclei for further unstructured chains. A recent study revealed a structural composition of the stored spider silk proteins typical for natively unfolded proteins, where the proteins adopt random coil or polyproline II-helical structures.[69] Such structure would indeed allow a rapid and energetically favoured conversion to β-pleated sheets. Supporting this theory, nanofibrils similar to amyloid fibrils have been described at the transition of the B-zone into the duct.[16] In this model, a structural reconfiguration after fibril formation would be necessary to yield a parallel-β structure, probably through shear forces and stretching (see earlier in text). However, the process of fibril formation as known from various amyloid structures is a quite slow process. Therefore, it seems very unlikely that the rapid thread formation during spinning is consistent with such an assembly pathway.

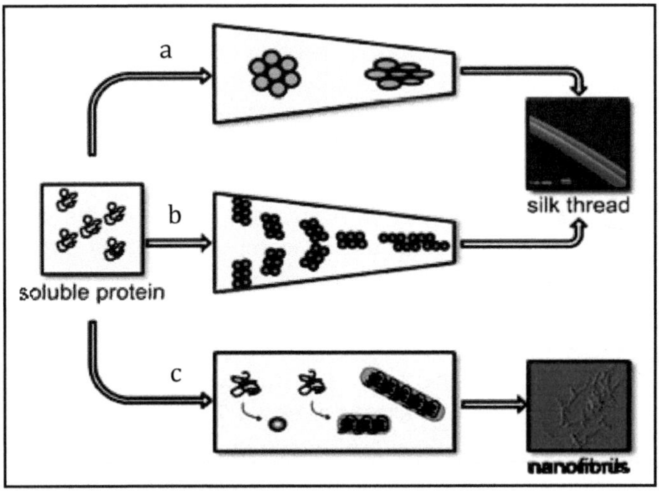

Figure 5.4 Potential assembly pathways towards silk thread formation (a and b) and self-assembly resulting in amyloid-like fibrils (c). (a) The micelle theory;[17,19,61,63] (b) the liquid-crystal spinning theory proposed by Vollrath;[4] (c) the nucleation-aggregation mechanism leading to the formation of amyloid fibrils.[21] Figure adapted from Slotta.[73]

5.3 Conclusion

The amyloid fold is thought to be a very ancient structure[70], very likely not restricted to specific proteins. Moreover, it is suggested to be an intrinsic property of polypeptide backbones in general, with hydrogen bonds being the predominant driving force for forming amyloid structures.[71] Nature might have taken advantage of this highly stable fold, and proteins evolved adopting this structure for specific structural and mechanical functions.[71,72]

Likewise, silks reveal a quite ancient structure, and silk structures probably have evolved several times independently.[5] Comparing silks and functional amyloids, similarities can be found concerning their structure-function relationships. Both exhibit a low sequence complexity of the respective proteins and a trend towards repetitive sequence compositions allowing a high conformational flexibility without a strict prevalence for a specific conformation. All amyloids exhibit a cross-β structure when assembled into fibrous structures,[71] while silks in contrast can feature different folds.[5] Strikingly, those silks with α-helical or cross-β structure are extruded or deposited, while the spinning process (as described earlier) leads exclusively to parallel-β silks.[5] Cross-β silks can be converted into parallel-β structures upon stretching,[33] indicating that mechanical stress and shear forces are crucial events leading to the parallel-β arrangement.

It can be speculated that "ancient" silk proteins primarily adopted an amyloid-like structure which can still be observed in nanofibrils as found in the B-zone of the major ampullate gland[16] and *in vitro* in fibrils of engineered spider silk proteins.[21] On the basis of the amyloid structure, spiders probably improved the spinning process to adopt the mechanical properties of silks according to special needs. Instead of forming a network of short amyloid fibrils with limited length, spiders produce long and continuous threads. The embedding of β-sheet crystals in an amorphous matrix within these threads leads to elaborate mechanical features combining the high strength provided by densely packed β-sheets with elastic properties. The highly complex process of stretching and aligning of silk proteins inside the spinning duct seems to lead unavoidably to a transition of any cross-β structure to the observed parallel-β structure which is found in all spun silk.

Acknowledgement

We thank Markus Heim for critical comments on our manuscript. This work was supported by the *Deutsche Forschungsgemeinschaft*.

References

1. Dobson, C. M. (2001) *Philos. Trans. R. Soc. Lond. B Biol. Sci.*, **356**, 133.
2. Smith, A. M., and Scheibel, T. (2010) *Macromol. Chem. Phys.*, **211**, 127.

3. Romer, L., and Scheibel, T. (2008) in *Fibrous Proteins*, (ed, Scheibel, T.), Landes Bioscience, Austin, p. 121.
4. Vollrath, F., and Knight, D. P. (2001) *Nature*, **410**, 541.
5. Craig, C. L. (1997) *Annu. Rev. Entomol.*, **42**, 231.
6. Hepburn, H. D., Chandler, H. R., and Davidoff, M. R. (1979) *Insect Biochem.*, **9**, 69.
7. Warwicker, J. O. (1960) *J. Mol. Biol.*, **2**, 350.
8. Gosline, J. M., Guerette, P. A., Ortlepp, C. S., and Savage, K. N. (1999) *J. Exp. Biol.*, **202**, 3295.
9. Gatesy, J., Hayashi, C., Motriuk, D., Woods, J., and Lewis, R. (2001) *Science*, **291**, 2603.
10. Hinman, M. B., and Lewis, R. V. (1992) *J. Biol. Chem.*, **267**, 19320.
11. Guerette, P. A., Ginzinger, D. G., Weber, B. H., and Gosline, J. M. (1996) *Science*, **272**, 112.
12. Winkler, S., and Kaplan, D. L. (2000) *J. Biotechnol.*, **74**, 85.
13. Vollrath, F., and Knight, D. P. (1999) *Int. J. Biol. Macromol.*, **24**, 243.
14. Heim, M., Keerl, D., and Scheibel, T. (2009) *Angew. Chem. Int. Ed.*, **48**, 3584.
15. Hardy, J. G., Romer, L. M., and Scheibel, T. R. (2008) *Polymer*, **49**, 4309.
16. Kenney, J. M., Knight, D., Wise, M. J., and Vollrath, F. (2002) *Eur. J. Biochem.*, **269**, 4159.
17. Hagn, F., Eisoldt, L., Hardy, J. G., Vendrely, C., Coles, M., Scheibel, T., and Kessler, H. (2010) *Nature*, **465**, 239.
18. Askarieh, G., Hedhammar, M., Nordling, K., Saenz, A., Casals, C., Rising, A., Johansson, J., and Knight, S. D. (2010) *Nature*, **465**, 236.
19. Eisoldt, L., Hardy, J. G., Heim, M., and Scheibel, T. R. (2010) *J. Struct. Biol.*, **170**, 413.
20. Hagn, F., Thamm, C., Scheibel, T., and Kessler, H. (2011) *Angew. Chem. Int. Ed.*, **50**, 310.
21. Slotta, U., Hess, S., Spiess, K., Stromer, T., Serpell, L., and Scheibel, T. (2007) *Macromol. Biosci.*, **7**, 183.
22. Knight, D. P., Knight, M. M., and Vollrath, F. (2000) *Int. J. Biol. Macromol.*, **27**, 205.
23. Nilsson, M. R., and Raleigh, D. P. (1999) *J. Mol. Biol.*, **294**, 1375.
24. Rocken, C., Ernst, J., Hund, E., Michels, H., Perz, J., Saeger, W., Sezer, O., Spuler, S., Willig, F., and Schmidt, H. H. (2006) *Dtsch. Med. Wochenschr.*, **131**, S45.

25. DeLellis, R. L., and Bowling, M. C. (1970) *Hum. Pathol.*, **1**, 655.
26. Turnell, W. G., and Finch, J. T. (1992) *J. Mol. Biol.*, **227**, 1205.
27. Khurana, R., Uversky, V. N., Nielsen, L., and Fink, A. L. (2001) *J. Biol. Chem.*, **276**, 22715.
28. Scheibel, T., and Serpell, L. (2005) in *Protein Folding Handbook. Part II*, (ed. Buchner, J., and Kiefhaber, T.), WILEY-VCH Verlag GmbH Co KGaA, Weinheim, p. 197.
29. Fraser, R. D. B., and MacRae, T. P. (1973) *Conformation in Fibrous Proteins*, Academic Press, Inc., New York.
30. Lefevre, T., Rousseau, M. E., and Pezolet, M. (2007) *Biophys. J.*, **92**, 2885.
31. Dong, Z., Lewis, R. V., and Middaugh, C. R. (1991) *Arch. Biochem. Biophys.*, **284**, 53.
32. Zandomeneghi, G., Krebs, M. R., McCammon, M. G., and Fandrich, M. (2004) *Protein Sci.*, **13**, 3314.
33. Geddes, A. J., Parker, K. D., Atkins, E. D., and Beighton, E. (1968) *J. Mol. Biol.*, **32**, 343.
34. Kajava, A. V., Squire, J. M., and Parry, D. A. (2006) *Adv. Protein. Chem.*, **73**, 1.
35. Sunde, M., Serpell, L. C., Bartlam, M., Fraser, P. E., Pepys, M. B., and Blake, C. C. (1997) *J. Mol. Biol.*, **273**, 729.
36. Parkhe, A. D., Seeley, S. K., Gardner, K., Thompson, L., and Lewis, R. V. (1997) *J. Mol. Recognit.*, **10**, 1.
37. Lucas, F., Shaw, J. T., and Smith, S. G. (1960) *J. Mol. Biol.*, **2**, 339.
38. Takahashi, Y., Gehoh, M., and Yuzuriha, K. (1999) *Int. J. Biol. Macromol.*, **24**, 127.
39. Asakura, T., Yao, J., Yamane, T., Umemura, K., and Ulrich, A. S. (2002) *J. Am. Chem. Soc.*, **124**, 8794.
40. Takahashi, Y. (1993) in *Silk Polymers* (ed. Kaplan, D. L., Adams, W., Farmer, B., and Viney, C.), American Chemical Society, Washington, DC, p. 168.
41. Lotz, B., and Colonna Cesari, F. (1979) *Biochimie*, **61**, 205.
42. Marsh, R. E., Corey, R. B., and Pauling, L. (1955) *Biochim. Biophys. Acta*, **16**, 1.
43. van Beek, J. D., Hess, S., Vollrath, F., and Meier, B. H. (2002) *Proc. Natl. Acad. Sci. U.S.A.*, **99**, 10266.
44. Simmons, A. H., Michal, C. A., and Jelinski, L. W. (1996) *Science*, **271**, 84.

45. Hayashi, C. Y., Shipley, N. H., and Lewis, R. V. (1999) *Int. J. Biol. Macromol.*, **24**, 271.
46. Roemer, L., and Scheibel, T. (2008) in *Fibrous Proteins* (ed. Scheibel, T.), Landes Bioscience, Austin, p. 137.
47. Zhou, C. Z., Confalonieri, F., Medina, N., Zivanovic, Y., Esnault, C., Yang, T., Jacquet, M., Janin, J., Duguet, M., Perasso, R., and Li, Z. G. (2000) *Nucleic Acids Res.*, **28**, 2413.
48. Papadopoulos, P., Solter, J., and Kremer, F. (2007) *Eur. Phys. J. E. Soft Matter*, **24**, 193.
49. Ene, R., Papadopoulos, P., and Kremer, F. (2009) *Soft Matter*, **5**, 4568.
50. Papadopoulos, P., Solter, J., and Kremer, F. (2009) *Colloid. Polym. Sci.*, **287**, 231.
51. Grubb, D. T., and Ji, G. (1999) *Int. J. Biol. Macromol.*, **24**, 203.
52. Scheibel, T. (2004) *Microb. Cell Fact.*, **3**, 14.
53. Papadopoulos, P., Ene, R., Weidner, I., and Kremer, F. (2009) *Macromol. Rapid Comm.*, **30**, 851.
54. Riekel, C., Madsen, B., Knight, D., and Vollrath, F. (2000) *Biomacromol*, **1**, 622.
55. Glisovic, A., Vehoff, T., Davies, R. J., and Salditt, T. (2008) *Macromol.* **41**, 390.
56. Liu, Y., Shao, Z., and Vollrath, F. (2005) *Nat. Mater.*, **4**, 901.
57. Rudall, K. M., and Kenchington, W. (1971) *Annu. Rev. Entomol.*, **16**, 73.
58. Dicko, C., Kenney, J. M., and Vollrath, F. (2006) *Adv. Protein Chem.*, **73**, 17.
59. Knight, D. P., and Vollrath, F. (2001) *Biomacromolecules*, **2**, 323.
60. Knight, D. P., and Vollrath, F. (2001) *Naturwissenschaften*, **88**, 179.
61. Rammensee, S., Slotta, U., Scheibel, T., and Bausch, A. R. (2008) *Proc. Natl. Acad. Sci. U.S.A.*, **105**, 6590.
62. Slotta, U. K., Rammensee, S., Gorb, S., and Scheibel, T. (2008) *Angew. Chem. Int. Ed.*, **47**, 4592.
63. Jin, H. J., and Kaplan, D. L. (2003) *Nature*, **424**, 1057.
64. Exler, J. H., Hummerich, D., and Scheibel, T. (2007) *Angew. Chem. Int. Ed.*, **46**, 3559.
65. Knight, D., and Vollrath, F. (1999) *Tissue Cell*, **31**, 617.
66. Hamley, I. W. (2010) *Soft Matter*, **6**, 1863.
67. Baron, M. (2001) *Pure Appl. Chem.*, **73**, 845.

68. Li, G., Zhou, P., Shao, Z., Xie, X., Chen, X., Wang, H., Chunyu, L., and Yu, T., (2001) *Eur. J. Biochem.*, **268**, 6600.

69. Lefevre, T., Leclerc, J., Rioux-Dube, J. F., Buffeteau, T., Paquin, M. C., Rousseau, M. E., Cloutier, I., Auger, M., Gagne, S. M., Boudreault, S., Cloutier, C., and Pezolet, M. (2007) *Biomacromolecules*, **8**, 2342.

70. Chernoff, Y. O. (2004) *Curr. Opin. Chem. Biol.*, **8**, 665.

71. Chiti, F., Dobson, C. M. (2006) *Annu. Rev. Biochem*, **75**, 333.

72. Fowler, D. M., Koulov, A. V., Balch, W. E., and Kelly, J. W. (2007) *Trends Biochem. Sci.*, **32**, 217.

73. Slotta, U. K. (2009) *Charakterisierung von Assemblierungsformen rekombinanter Spinnenseidenproteine.* PhD thesis,. Department Chemie, Institut für Organische Chemie und Biochemie, TU München.

Chapter 6

Yeast Prions: Their Assembly into Protein Fibrils and the Role of Assembly Modulators

Luc Bousset and Ronald Melki
Laboratoire d'Enzymologie et Biochimie Structurales,
Centre National de la Recherche Scientifique, Bâtiment 34, 1 Avenue de la Terrasse, 91198 Gif-sur-Yvette, France
melki@lebs.cnrs-gif.fr

Prions are proteins that propagate phenotypic traits through self-perpetuating folding events. The nature of the conformational change that conveys the functional constitutive form of a prion protein to a form with persistent altered function and the ability to convert constitutive prion molecules into modified prions is poorly understood. We present here what is known about the assembly into protein fibrils of four yeast prions in test tubes. We also describe how molecular chaperones modulate the assembly of yeast prions *in vitro* and through their assembly, promoting and sequestering activities, tune the propagation of prions phenotypic traits from mother to daughter cells. Finally the notion of prion "infectivity" is discussed in a cellular context and from a structural perspective.

The Functional Fold: Amyloid Structures in Nature
Edited by Suzi Jarvis and Anika Mostaert

ISBN 978-981-4267-40-3 (Hardcover), 978-981-4267-62-5 (eBook)
www.panstanford.com

6.1 Introduction

The transmission of defined visible traits, that is, phenotypes from one generation to the next, is a central aspect in living organisms, and the nucleic acid sequence units that constitute the genes are the stable entities of inheritance. While the appearance of new traits has long been considered as the sole consequence of mutations in nucleic acid sequences, studies from the past two decades have proven that new traits can be passed between individuals without changes in nucleic acid sequences.

The revolutionary idea that some proteins could behave as infectious agents carrying alone the hereditary information that ensures their propagation was first proposed by the mathematician Griffith.[1] Such proteins are now termed prions for "infectious proteins"[2] and their ability to transmit novel phenotypes is believed to be the consequence of their capacity to adopt more than one stable conformation. Indeed, since the structure of a polypeptide defines its function, cells containing the same polypeptide chain in different conformations will have distinct phenotypes. However, if newly synthesized prion molecules adopt in an independent manner a wide range of different conformations, all cells would exhibit the same phenotype reflecting an average functional state. In contrast, if the conformation of such proteins is constrained in different cells, distinctive phenotypes will be displayed.

6.2 Prions in Yeast

In the baker's yeast, *Saccharomyces cerevisiae*, three traits, [PSI$^+$], [URE3], and [PIN$^+$], and in the filamentous fungus, (*Podospora anserine*), one feature, [Het-s], are inherited in a non-Mendelian manner.[3–6] It was first thought that these traits were due to a non-chromosomal nucleic acid. However, [PSI$^+$], [URE3], [PIN$^+$], and [Het-s] inheritance differs significantly from that of DNA plasmids, RNA viruses, the mitochondrial genome, or RNA replicons. In addition, the genes encoding these traits are located in the nucleus of yeast cells.[6–8] This led to the idea that these traits are due to the prion properties of the proteins Sup35, Ure2, Rnq1, and HET-s, respectively.[9]

6.3 The [PSI+], [PIN], [URE3], and [Het-s] Phenotypes

6.3.1 The [PSI+] and [PIN] Traits

The [PSI+] trait is associated with Sup35p, an essential component of the translation termination machinery.[10,11] Indeed, the eukaryotic release factor 3 (eRF3) or Sup35p, constitutes together with eRF1 or Sup45p, the functional translation release factor that recognizes stop codons and releases nascent polypeptides from the final t-RNA.[12–14] Sup35p is soluble and functional in [psi−] cells, while it is in functionally compromised aggregated form in [PSI+] cells.[15,16] Thus, while translation termination is efficient in [psi−] cells, it is not in [PSI+] cells, and the ribosomes read-through stop codons and synthesize abnormal polypeptides.

Sup35p (Swiss-Prot P05453) is a large polypeptide made of 685 amino acid residues with a calculated molecular mass of 76.5 kDa.[17] Three structural/functional regions have been distinguished within Sup35p (Fig. 6.1). The N-terminal region (N) that extends from amino acid residues 1 to 122 is rich in Q, N, and G residues (47%). This region is not essential for the role of the protein in translation termination.[18–20] It is called prion domain (PrD) as it plays a critical role in prion propagation. The C-terminal region is the functional domain of the protein providing translation-termination activity.[13,18] The latter domain is compactly folded based on the crystal structure of eRF3 from *S. pombe*[21] and contains the GTP and eRF1-binding sites (Fig. 6.1; PDB accession code 1R5N). Finally, the middle region (M) extending from amino acid residues 123 to 253 links the PrD and the functional domain.

The overexpression of Sup35p increases the frequency of [PSI+] appearance. This is also what is observed upon the overexpression of Rnq1p (Swiss-Prot P25367), a 42.5 kDa polypeptide, with unknown function. Rnq1p is rich in Q and N residues particularly in its C-terminal part extending from amino acid residue 153 to 405 (Fig. 6.1). Rnq1p is probably a natively unfolded polypeptide. It exists as a soluble protein as well as in insoluble aggregates.[22] The amount of insoluble Rnq1p increases when the protein is overproduced.[23] In its insoluble state, Rnq1p allows the conversion of [psi−] cells into [PSI+] cells and is at the origin of the phenotype [PIN+] that comes from "[PSI+] inducibility".[5]

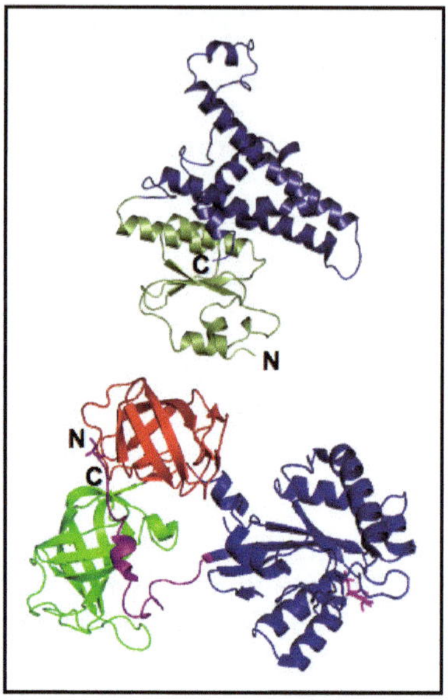

Figure 6.1 Primary structures of Sup35pNM, Ure2p1-93, Rnq1p153-405, and Het-S1-289. Q and N residues are coloured red, S and T are coloured blue. Residues constituting the domains required for prion propagation are in bold. The structure of the compactly folded N-terminal domains of Sup35p from *S. pombe* and that of Ure2p from *S. cerevisiae* are shown. The three sub-domains of Sup35p C-terminal domains are coloured blue, red, and green, and the peptide that links this domain to the M domain of the protein is coloured purple. The two sub-domains of Ure2p C-terminal domain are coloured blue and green. The letters N and C in the structure panels refer to the N- and C-terminal residues of the polypeptide represented. The GTP molecule bound to Sup35p is coloured purple. The structures were generated with PyMol (http://pymol.sourceforge.net/). The PDB coordinates used for the panels A and B are 1R5N and 1G6Y, respectively. See also Colour Insert.

6.3.2 The [URE3] Trait

Nitrogen catabolism is finely regulated in *S. cerevisiae*. Ure2p, for which the exact function is unknown, is involved in this process.[24–28] In its soluble functional state, Ure2p is believed to interact in the yeast cytosol with the "GATA" transcription factor Gln3p. Thus, Ure2p prevents the entry of Gln3p into the nucleus where it transcribes a number of genes among which includes *DAL5*, the gene encoding the allantoate transporter Dal5p.[29] In its functionally compromised form associated to the [URE3] trait,[30] Ure2p is not capable of sequestering Gln3p in the cytoplasm and the transcription of a number of genes becomes constitutive. For example, the cells express Dalp5p and import in a constitutive manner; ureidosuccinate, a precursor of uracil, only imported by the yeast cells when they are grown in the absence of simple nitrogen sources such as ammonium or glutamine.

Ure2p (Swiss-Prot P23202) has a calculated molecular mass of 40.2 kDa and is composed of 354 amino acid residues (Fig. 6.1). The flexible N-terminal part of Ure2p extending from amino acid residue 1 to 93[31] is very rich in Q, N, S, and T residues (62%). This region is required for the propagation of the [URE3] phenotype[32] and is referred to as the PrD of the protein. The C-terminal domain that extends from amino acid residues 94 to 354[31] complements *URE2* gene deletion and therefore constitutes the functional domain of the protein.[32] It is compactly folded (Fig. 6.1; PDB accession code 1G6Y),[33–35] binds glutathione (PDB accession code 1JZR),[36] and exhibits glutathione perroxidase activity.[37]

6.3.3 The [Het-s] Trait

The fusion of two different *Podospora anserina* filaments contacting each other into heterokaryon is tightly regulated and requires, among other things, that the two filaments are genetically identical at the *het-s* locus.[38] Two common alleles of this locus, called *het-s* and *het-S*, encode cytosolic polypeptides that differ in 13 amino acid residues out of 289 distributed throughout the protein.[39] The protein HET-s exist under two states termed [Het-s*] and [Het-s]. Strains in the [Het-s*] state are indifferent to the *het-s*/*het-S* status of their fusion partner. In contrast, an apoptotic reaction occurs leading to the death of the heterocaryotic cell upon fusion of a filament in its [Het-s] state with a filament containing the protein HET-S.[6] The HET-s protein (TrEMBL Q03689) has a calculated molecular mass of 32 kDa. It has no remarkable features (Fig. 6.1). The C-terminal domain of the protein spanning amino acid residues 218–289 is nowadays considered essential for prion propagation, although the N-terminal 26 amino acid residues and point mutations at amino acid residues 23 and 33 were previously thought to be at one point as critical for prion propagation.[40,41]

6.4 *In vitro* Assembly of Sup35p, Rnq1p, Ure2p, and HET-s into Protein Fibrils

Purified full-length Sup35p, Rnq1p, and Ure2p are soluble and have a substantial helical content. In contrast, HET-s is poorly soluble and is purified under denaturing conditions. Sup35p, Rnq1p, and Ure2p assemble at neutral pH and under physiological salt concentrations into high molecular weight oligomers. These oligomers are either soluble and spherical or insoluble and fibrillar (Fig. 6.2).[31,42–45] The formation of the unbranched fibrils is greatly accelerated by seeding with preformed fibrils suggesting that the limiting step in the assembly process is nucleation. Fibrils also form upon dilution from denaturant of HET-s.[46]

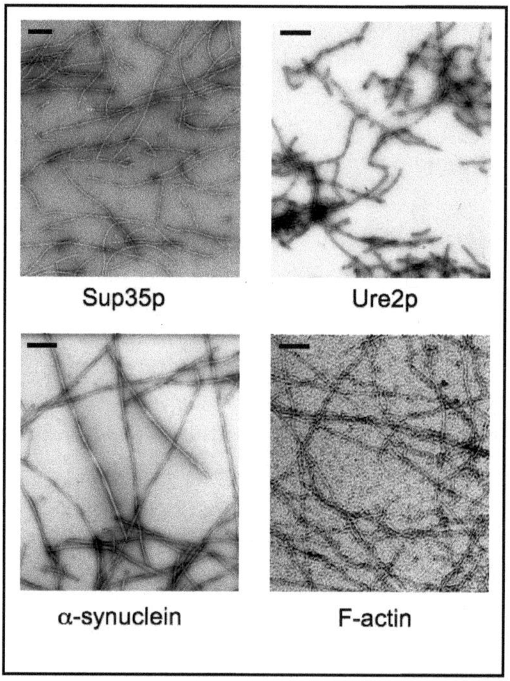

Figure 6.2 Electron micrographs of negatively stained samples of fibrillar Sup35p, Ure2p, α-synuclein, and actin. Bar = 0.2 μm.

Fibrils formed *in vitro* from native full-length Sup35p and Ure2p (Fig. 6.2) resemble canonic amyloids such as those made of α-synuclein (Fig. 6.2). They also resemble native assemblies such as filamentous actin (Fig. 6.2). Indeed, Sup35p and Ure2p fibrils are unbranched, approximately 20 nm wide and more than 1 μm long,[31,43–44] and have increased resistance to proteolysis when compared to the soluble proteins.[31,43] Similar to conventional amyloids, fibrillar Sup35p and Ure2p bind the dyes Congo red and thioflavin T, and exhibit yellow-green birefringence in polarized light upon Congo red binding.[43,47] These characteristics, together with the findings that the N-terminal domains of Sup35p and Ure2p that are critical for prion propagation, assemble *in vitro* into fibrils that exhibit a 4.7 Å reflection in X-fibre diffraction images,[48,49] led to the view that full-length Sup35p and Ure2p fibrils are conventional amyloids. This view needs to be tempered based on the observations summarized below.

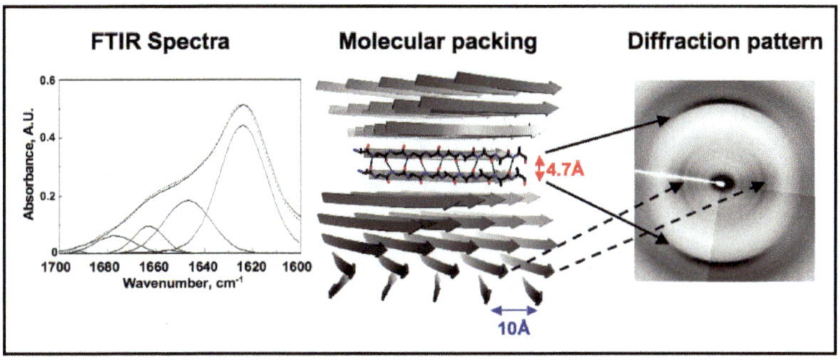

Figure 6.3 Amyloid fibrils structural characteristics. FTIR spectrum of amyloid fibrils (left panel) showing the amide I absorbance maximum at 1610–1630 cm^{-1}. The β-sheet component of the fibrils absorbs specifically within this wavelength range unlike the sheet content of soluble proteins. Packing of the β-strands within an amyloid fibril (middle panel). The systematically H-bonded (dashed lines) β-sheet core is represented. Oxygen atoms are in red while nitrogen atoms are in blue. X-ray diffraction pattern of amyloid fibrils (right panel). As indicated with the arrows, the 4.7 Å and 10 Å reflections originate from the interstrand and intersheet distances, respectively. The orientation of the strands and sheets relative to the fibril main axis that is perpendicular to the scheme is shown. See also Colour Insert.

It is worth recapping first the structural definition of amyloid fibrils. The term amyloid fibrils refers to fibrillar protein deposits associated with disease that appear unbranched; that bind the dyes thioflavin T or S and Congo red, with a typical and concomitant birefringence in polarized light in the latter case; that have a high content of β-structures in a cross-β arrangement, where the hydrogen bonds between two consecutive sheets are oriented parallel to the main fibril axis while the constituting β-strands are oriented transversely to the main fibril axis.[50,51] This type of strongly hydrogen-bonded structure (Fig. 6.3) that absorbs specifically infrared light between 1618 and 1623 cm^{-1},[52] gives rise to a characteristic pattern of reflections in X-ray diffraction experiments. This pattern consists of a conserved 4.6 to 4.8 Å meridional spacing and an equatorial spacing of approximately 10 Å (Fig. 6.3). The 4.6 to 4.8 Å reflection comes from the distance between two hydrogen-bonded strands, and is invariant as it depends on the geometry of the polypeptide backbone. It is referred to as the "main chain spacing". The equatorial reflection at approximately 10 Å comes from the packing distance between two juxtaposed β-sheets.[53,54] This distance can vary with the amino acid composition of the polypeptide as it depends on the orthogonal protrusion of the amino acid side chains from the plane of the sheet. It is worth noting that this reflection is not observed when the intersheet spacing is not regular.

Finally, as the sheets forming the backbone of the fibrils are systematically hydrogen bonded (Fig. 6.3) and are poorly exposed to the solvent, the backbone amide hydrogen atoms within amyloid fibrils are extremely resistant to hydrogen/deuterium exchange.[55–62]

While full-length Ure2p fibrils have been extensively characterized structurally, full-length Sup35p fibrils have not. The following section will therefore develop what we learned from the structural characterization of fibrillar Ure2p without precluding what a thorough structural characterization of full-length Sup35p fibrils might reveal. As indicated above, fibrillar Ure2p has increased resistance to limited proteolytic treatments; however, the degradation patterns of the soluble and fibrillar forms of the protein are very similar,[32,63] suggesting that the assembly of Ure2p into protein fibrils is not accompanied by a major conformational change. The absence of conformational changes within the compactly folded helical C-terminal domain of Ure2p upon assembly, of full-length Ure2p into fibrils, is further supported by the finding that the protein binds glutathione,[36] and exhibits glutathione perroxidase activity[37] within the fibrils as does the native, soluble form.

Two methods ideally suited for analyzing the β-sheet content of macromolecular assemblies further suggest that Ure2p fibrils are not classical amyloids. Indeed, Fourier-transform infrared spectroscopy (FTIR) reveals a large proportion of α-helical structure within the fibrils, very limited structural change upon assembly of soluble Ure2p into fibrils, and the absence of increased absorbance in the amyloid specific band (1618–1623 cm^{-1}).[42] Further, X-ray fibre diffraction patterns generated from partially aligned Ure2p fibrils are inconsistent with a cross-β structure and compatible instead with the regular packing of globular Ure2p molecules.[64] Finally, the findings that (i) fibrils made of full-length Ure2p exhibit a punctate appearance on averaged high-resolution cryoelectron micrographs only compatible with a highly repetitive structure made up of ordered arrangement of globular Ure2p molecules stacked in a helical array[65] and (ii) Ure2p molecules are not involved in a systematically H-bonded β-sheet structure within the fibrils as determined using hydrogen/deuterium exchange studies and mass spectrometry,[66] strongly suggest that Ure2p fibrils are not conventional amyloids.

Fibrils made following dilution from denaturant of unfolded HET-s are of amyloid nature based on their FTIR absorbance spectrum.[46] It is, however, not clear yet whether authentic HET-s forms such fibrils.

6.5 Structural Basis of Yeast Prion Propagation

It is widely believed that prions propagate through unfolding or misfolding events. Two mechanistic models have been proposed to illustrate how following this seminal folding event prions propagate: the template assistance[1,67,68] and the seeded-polymerization models.[69]

The infectious misfolded form interacts physically with the native soluble form and transforms the latter into a misfolded form following the template-assistance model (Fig. 6.4a). As the infectious homodimer/polymer is capable of dissociating/breaking, the conversion process of the native pool of prion into misfolded forms is exponential. In the template-assistance model, the infectious form acts somehow as a template facilitating the occurrence of an energetically unfavourable protein-folding event. A variant of this scenario has been proposed where the interaction between an unidentified enzyme or chaperone and native prion proteins could favour the protein folding event at the origin of the prion form.[70]

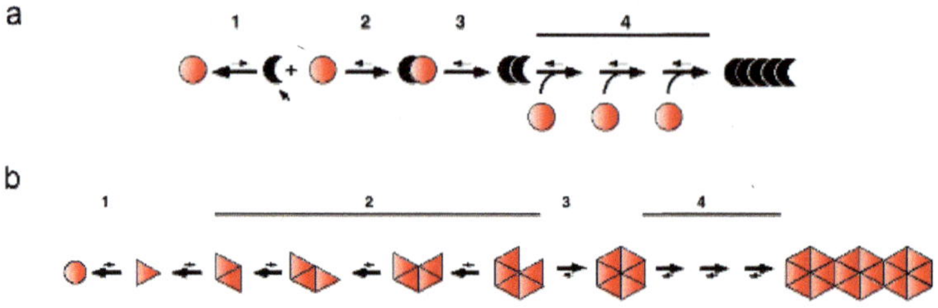

Figure 6.4 Theoretical models for prion propagation. (a) Template assistance model. (1) The constitutive form of the prion protein undergoes a conformational change that leads to an abnormal form. This event either occurs at a low frequency, where the abnormal form is unstable, or is highly sensitive to clearance. Thus, the concentration of the abnormal form is low. (2) The prion protein in its abnormal form interacts with the constitutive form. (3) The constitutive form of the protein is converted into an abnormal form through this interaction. (4) The abnormal homodimer can grow indefinitely by incorporating additional prion molecules following their conversion into an abnormal form. It can also fragment. Each fragment is capable of growing by incorporation of converted prion molecules. (b) Seeded-polymerization model. (1) Following a seminal conformational change, the abnormal form of the prion protein acquires the ability to interact with another prion molecule in an abnormal form. The interaction is unstable because the intermolecular interactions are not strong enough to outweigh the entropic cost of binding. (2) Thus, the low molecular weight oligomers that are formed dissociate until a stable nucleus is formed. (3) This nucleus or seed can grow indefinitely from one or both ends depending on the structural properties of the abnormal form of prion protein. (4) It can also break into smaller stable fibrils that can elongate by incorporation of the abnormal form of the prion protein. The monomeric and oligomeric forms of the abnormal prion can be of exogenous origin in iatrogenic or dietary infections in vertebrates. See also Colour Insert.

The native monomeric prion molecule P is in equilibrium with a rare and unstable conformational isoform P^* according to the seeded-polymerization model (Fig. 6.4b). Because P^* is unstable, or highly sensitive to clearance, its concentration is very low. However, unlike P, P^* can interact with another P^* molecule. Similar to a classical crystallization event, it is thought that the formation of low molecular weight oligomers of P^* are not energetically favourable until a stable nucleus P^n, that can grow by incorporation of P^* at its ends, is formed as the energy gained from intermolecular interactions does not outweigh the entropic cost of binding. A number of inherited mutations that destabilize yeast prions predispose them to convert to their unstable P^* form,[71,72] thus, increasing the concentration of the latter form and favouring their oligomerization. The requirement for stable nuclei to form before conversion is stable accounts for the low frequency of occurrence of the prion phenotypes. However, the high efficiency of incorporation of P^* into oligomers and polymers made of P^* and breakage of the polymers into smaller units, each being a potential seed, accounts for the efficiency of transmission from mother to daughter cells.

6.6 Structural Models at an Atomic Resolution for Fibrillar Prion Proteins

High-resolution three-dimensional structures for prion proteins, in particular yeast prions, are not yet available. A number of models have been built mainly from low-resolution structural information obtained from fragments of the proteins and by analogy to high-resolution structural data obtained for short peptides (7–15 amino acid residues long) that do not necessarily have prion properties.[73–75] These working models should be manipulated with extreme caution as they may not reflect the authentic structures of the fibrillar form of full-length prion proteins. Nevertheless, biochemical and cellular data provide interesting information on the packing of prion molecules within the fibrils they form.

As mentioned previously, prion proteins are peculiar in that the same polypeptide chain can misfold into structurally distinct protein aggregates at the origin of distinguishable phenotypes named "prion strains" or "variants".[20,76–78] A set of data based on side-chain modification of amino acid residues located within the N-terminal domain of Sup35p NM fragment mutational analysis, cross-linking, NMR, and peptide arrays studies suggest that sub-regions within the region spanning residues 1–121 form the core of Sup35pNM fibrils. The length of these sub-regions is still subject to debate. Indeed, while data suggest that a significant portion of residues in the region spanning residues 1–121, residues 31–86, and 21–121[79] establish intermolecular contacts and form the core of distinct fibrillar assemblies that dictate the prion strain, other data suggest that the establishment of intermolecular interactions between short sequences within the region 1–121, residues 9–30, 59–86, and 90–130[80] govern the assembly of Sup35p into structurally distinct high molecular weight protein oligomers at the origin of prion strains variation.

In the case of Ure2p, real-time fluorescence microscopy observations have revealed that the fibrils are polar structures.[81] This somehow suggests that the fibrillar scaffold is complex and that Ure2p molecules are not stacked simply in-register in a cross-β structure. Indeed, if the molecules were stacked as in a cross-β structure, the two ends of the fibrils would be identical and the fibrils would not exhibit polarity. In addition, a number of biochemical indications on the molecular packing within the fibrils are available. Degradation studies strongly suggest that both the N- and C-terminal domains of the protein are involved in the fibrillar scaffold.[63] Furthermore, cross-linking studies indicate that the N- and C-terminal domains of Ure2p are associated through intramolecular interactions within monomeric Ure2p,[82] and have led to the identification of one of the surface interfaces involved in intermolecular interactions within the fibrils.[83]

6.7 Molecular Chaperones and the Propagation of Yeast Prions

Prion phenotypes can disappear spontaneously in yeast with a low frequency ($\sim 10^{-6}$) comparable to chromosomal gene mutations.[84] The frequency with which a prion phenotype is lost upon cell division defines the stability of a prion "strain" or "variant". When yeast prion "strains" were first observed, a difference in their propagation efficiency upon overexpression of Hsp104 was described.[20,85] The molecular chaperone Hsp104 is strictly required for yeast prion propagation as strains carrying *HSP104* gene deletion are unable to propagate the [PSI$^+$] and the [URE3] phenotypes.[85,86] Hsp104 is not the only cellular factor necessary for the continued propagation of prions in yeast cells. Other molecular chaperones play critical roles.[87] Indeed, the expression levels of Hsp70 family members have been reported to be important for [PSI$^+$] and [URE3] stability,[88–90] and mutations in both the peptide or the ATP-binding site of a number of Hsp70 family members abolish [URE3] propagation.[91,92] Finally, [URE3] and [PIN⁺] faithful propagation is dependent on the overexpression levels of members of the hsp40 family as [URE3] is lost upon the overexpression of Ydj1p,[86,88] while [PIN⁺] propagation requires the expression of Sis1p.[93]

A number of models have been proposed to account for the role played by molecular chaperones in the maintenance or the destabilization of the prion phenotypes. In one model, the elevated levels of molecular chaperones facilitate the disaggregation and remodelling of the high molecular weight species of prion proteins that act as seeds.[94,95] Such a remodelling activity may generate sites for the incorporation of additional prion proteins and favour prion protein aggregation. In another model, molecular chaperones disfavour assembly by sequestering either the folding intermediate(s) that assemble into prion aggregates, or the cellular factor(s) that are required for the generation of the high molecular weight species of prion proteins that act as seeds.[90] Molecular chaperones have also been attributed with a mechanistic role in the propagation from mother to daughter cells of Sup35p and Ure2p high molecular weight aggregates associated with [PSI⁺] and [URE3].[96]

Additional roles for molecular chaperones can be proposed. Molecular chaperones could act on prion protein conformers and modulate prion aggregation either by refolding or unfolding conformers that are aggregation prone, or by facilitating prion folding into aggregation prone states. In the first case, molecular chaperones will have an unfavourable effect on oligomerization, while in the second, they will favour the aggregation reaction. Molecular chaperones could also act as a matrix, imprinting a structure to prion oligomeric species, thus mimicking a seed and favouring assembly. Similarly, assembly could be facilitated by the interaction of several prion molecules with a single molecular chaperone as the latter becomes a sort of local concentrator facilitating the thermodynamically unfavourable nucleation process. Finally, molecular chaperones could restrain prion aggregation either by pacifying elongation sites, favouring the bundling of prion aggregates, thus reducing the exposure to the solvent of elongation sites, or sequester assembly competent prion conformers into structured or amorphous aggregates (Fig. 6.5).

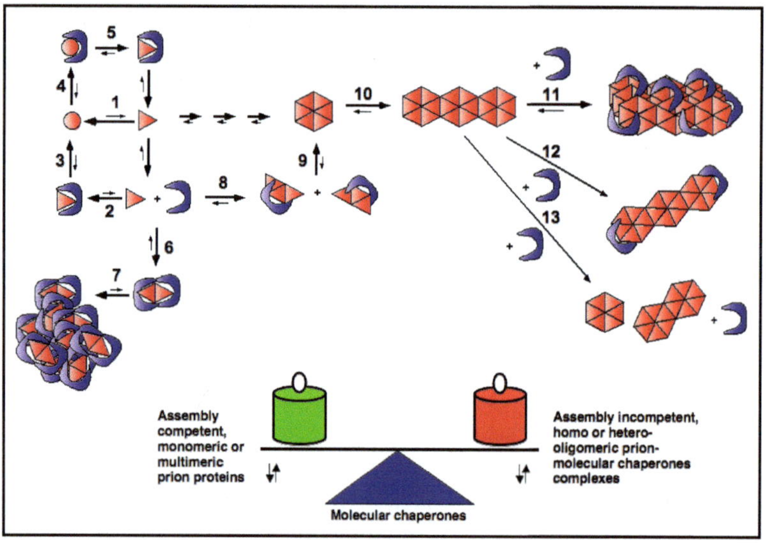

Figure 6.5 Modulation of prion aggregation by molecular chaperones. Molecular chaperones (in blue) can interact either with the abnormal form of the prion protein (red triangle, reaction 2) generated by a folding event (reaction 1), or the constitutive form of the prion (red circle, reaction 4). The interaction of molecular chaperones with the abnormal form of the prion can lead to the refolding of this form into the constitutive form (reaction 3). The interaction of molecular chaperones with the constitutive prion can lead to the generation of an abnormal form of the prion (reaction 5). Following the interaction of molecular chaperones with the abnormal, assembly competent form of the prion, an assembly incompetent low (reaction 6) or high (reaction 7) molecular weight complex can be generated. Molecular chaperones can also interact with an assembly competent oligomeric form of prions (reaction 8) and favour stable nuclei formation (reaction 9). Finally, molecular chaperones can interact with the walls of fibrillar prions and lead to the formation of fibrillar bundles (reaction 11) or the capping of fibril ends (reaction 12), or they can fragment the fibrils (reaction 13). Reactions 4, 5, 8, 9, and 13 favour prion assembly, while reactions 2, 3, 6, 7, 11 disfavour assembly. Thus, the functional differences between molecular chaperone actions can modulate the propagation of prion traits through a fine tuning of the oligomeric state of prion proteins. See also Colour Insert.

Four factors contribute to the extreme difficulty in assessing the role of molecular chaperones from observations made *in vivo.*

In vivo, the effect of a molecular chaperone can be counterbalanced by another chaperone. In addition, it is difficult to determine, with accuracy, the concentration of the different molecular chaperones and their variation throughout the cell cycle. Finally, not only is the activity of molecular chaperones modulated by cofactors for which concentration and activity vary over time, but their "specific activity", that is, the fraction of molecular chaperone A, B, or C that is available to interact with prions, is also impossible to determine. Given that it is very difficult to assess the exact role of each molecular chaperone in prion propagation from *in vivo* studies, a number *in vitro* studies were carried out to document the exact role of each molecular chaperone in yeast prion assembly. Such studies are limited by the variety of molecular chaperones to document, on the one hand, by the number of molecular chaperone combinations one can make, on the other.

Hsp104p was shown to reduce the nucleation phase preceding assembly and strongly stimulate the apparent assembly rate of full-length Sup35p. This effect could either be accounted for by an Hsp104p-mediated remodelling of the fibrillar form of Sup35p, for example, by fragmentation, with the generation of large number of nuclei, or by an Hsp104p-mediated nucleation activity. One study reported the active and GTP-mediated fragmentation of Sup35p by Hsp104p,[97] while another reported no such activity, but instead that Hsp104p favours *de novo* Sup35p oligomerization and nucleation.[98] Despite the lack of agreement of the Hsp104p mechanistic action, both studies strongly suggest that the greater segregation efficiency of the infectious Sup35p high molecular weight oligomers between mother and daughter cells is the consequence of an Hsp104p-mediated generation of Sup35p polymers that are smaller than those that form in the absence or upon inactivation of Hsp104p.

Similar to what has been described for Sup35p, Hsp104p has been reported to sever preformed Ure2p fibrils,[97] or not.[99] In addition, while Sup35p assemblies are of fibrillar nature in the presence of Hsp104p, Ure2p high molecular weight aggregates are not.[98,99] They appear amorphous in the electron microscope, and unlike Ure2p fibrils are SDS-sensitive.[99] Hsp70 family members alone or in cooperation with their co-chaperones from the Hsp40 family sequester Sup35p and Ure2p in an assembly incompetent state.[98,99] Finally, members of the Hsp40 family interact with Sup35p and Ure2p and disfavour assembly into fibrils.[98,99]

The functional interplay between Hsp104p and the Hsp70-40 systems and the consequences of changes in the cellular levels of molecular chaperones on the assembly of Sup35p and Ure2p into high molecular weight oligomers, in particular fibrils, were also documented *in vitro* by varying the concentrations of molecular chaperones in Sup35p and Ure2p assembly reactions and measuring the affinity of molecular chaperones for the prion proteins. Hsp70/40 possess higher affinities for Sup35p and Ure2p than Hsp104p, and were also shown to counteract the assembly stimulatory effect of Hsp104p.[98,99] The differences in affinities of the Hsp70/40 system and Hsp104p for yeast prions, and their sequestering and assembly promoting activities, suggest that molecular chaperones expression levels in a dividing yeast cell and variations in these levels throughout the cell cycle, or when the cells are subject to stress, finely tune the propagation and cure of the [PSI$^+$] and [*URE3*] prion traits (Fig. 6.5).

6.8 Nature of the Infectious Form of Yeast Prions

The fibrillar form of Sup35p, Ure2p, Rnq1p, and HET-s is widely believed to be the infectious form of these prion proteins, responsible for the emergence and propagation of the prion phenotypes [PSI$^+$], [PIN$^+$], [URE3], and [Het-s]. Indeed, an ultra-structural analysis suggested that Ure2p might be fibrillar in [URE3] cells.[100] In addition, fibrils made *in vitro* from recombinant Sup35p, Ure2p, Rnq1p, and HET-s induce the *de novo* appearance of the prion phenotypes when re-introduced in yeast or cells of filamentous fungi.[101–105] However, it is important to bear in mind that the fibrillar form of prions are in equilibrium with soluble oligomeric forms that are introduced within the cells at the same time as the fibrillar form. It is therefore unclear whether the fibrillar form is the only infectious form and the specific infectivity of the different oligomeric forms remains unknown.

Nucleated assembly is the simplest molecular process that can account for the propagation of a prion phenotype upon introduction within a cell of a mixture of fibrillar and oligomeric non-fibrillar forms of the recombinant prion. However, one can envisage that prion phenotype emergence and propagation upon introduction within a cell of oligomeric forms of a recombinant prion is the consequence of other events. Indeed, *in vivo* and *in vitro* studies of the role of molecular chaperones in the stability of prion phenotypes, and the aggregation of yeast prions, have shown that molecular chaperones bind and remodel prion high molecular weight oligomeric forms. They have also shown that the levels of molecular chaperones tune, in a very tight manner, the aggregation of prions, and certainly the emergence of the prion phenotypes. It is therefore reasonable to envisage that the reintroduction of a high molecular weight oligomeric forms of a prion within a cell imbalances the molecular chaperone proteostasis allowing the *de novo* formation of an infectious form. One could argue that this cannot account for the strain-specific phenotypes and indeed, one could easily imagine that fibrils and soluble high molecular weight oligomers possessing different quaternary structures; in other words, those that expose a different surface area to the solvent, interact with different molecular chaperones. As a consequence, the functional pool of chaperones can vary in cells where aggregates of different types have been introduced, and "strain-specific" prion high molecular weight aggregates can form and propagate *de novo*. Thus, the emergence and propagation of a particular yeast prion strain could be the combined consequence of a specific change in the functional pool of molecular chaperones, on the one hand, and a nucleated assembly process, on the other. Yeast prions do not spread from cell to cell. They are stably inherited by daughter cells from mothers or passed between partners during mating. To better assess the potential role of molecular chaperones in the *de novo* formation of prion particles and the propagation of prion phenotypes it is critical to (i) document the changes in molecular chaperone functional pool and expression profile upon introduction of recombinant fibrillar prion within cells, (ii) determine whether the molecular chaperone-binding pattern of different types of fibrillar prions differ, and how, by a proteomic approach.

The assembly of yeast prions is often considered as being driven by polyQ/N extensions. This is certainly not the case for HET-s as the latter protein lacks Q and N residues in its PrD(s). In addition, the increase of Q and N proportion in the N-terminal domain of Ure2p from 46% (in wild-type Ure2p) to 54% (in a variant Ure2p) is accompanied by a marked reduction in assembly.[106] This clearly indicates that assembly is not solely driven by the Q/N extensions of yeast prions.

6.9 Conclusions and Outlook

It is worth concluding this chapter with the central question: What makes, by analogy to PrP aggregates, yeast prions "infectious" and are other protein aggregates involved in human diseases, that are similar in shape or by their secondary structure content, potentially "infectious"? A partial answer to this question comes from the observation that the vast majority of Ure2p and Sup35p homologues that are expressed in various yeast strains and that possess Q/N extensions do not, apart from a unique exception, induce [URE3] or [PSI$^+$] phenotypes in their respective yeast strains.[107] In contrast, a number of these homologues induce the prion phenotype when expressed in *S. cerevisiae*.[108,109] This suggests that *S. cerevisiae* is particularly suitable for, and sensitive to, infectious proteins emergence and propagation, and might reflect a particular molecular chaperone specific proteostasis in this yeast strain. The only protein aggregates associated with disease, that are considered as infectious, are those containing PrP. Recently, however, the cytoplasmic protein aggregates enriched in fibrillar α-synuclein, known as Lewy bodies, that are associated with Parkinson's disease, were shown to form in healthy human fetal midbrain tissue transplanted into the brains of Parkinson's diseased patients.[110] This observation suggests that the aggregated α-synuclein, released from diseased neurons are capable of "infecting" neighbouring healthy tissue. Similarly, other protein aggregates associated with degenerative diseases such as Amyloid-β, huntingtin, β2-microglobulin, and IAPP could play the role of vectors in the formation of cytotoxic extracellular protein deposits associated with Alzheimer's and Huntington's diseases, dialysis-related and systemic amyloidosis propagation.

If this is indeed the case, then a wide variety of protein deposits associated with disease would be infectious, that is, have prion properties and could, through their aggregated state, propagate phenotypes/diseases. What would distinguish these protein aggregates from those we consider nowadays as prions is either their capacity to recruit soluble polypeptide chains or their intrinsic structural stability. Indeed, if the recruitment of the soluble form of polypeptide A is less efficient by protein aggregate A than that of a polypeptide B by protein aggregate B, aggregate A will grow slower than B and will distribute less efficiently from mother to daughter cells with, as a consequence, its loss during cell divisions. Similarly, if the intrinsic structural stability of protein aggregate A, which is reflected in the ability of this protein aggregate to fragment or be fragmented, is higher than that of protein aggregate B, aggregate A will grow indefinitely without generating smaller aggregates. As a consequence, this protein aggregate will not distribute evenly between mother and daughter cells and will be lost in contrast with protein aggregates that remodel with ease. Finally, if the structural characteristics of protein aggregate A allow its clearance by the cellular degradation machinery while those of aggregate B do not, aggregate A will propagate less efficiently than aggregate B. Thus, to understand what makes prions "infectious" it is critical to determine at an atomic resolution their structural characteristics and compare them to that of protein aggregates involved in disease.

Acknowledgement

This work was supported by the CNRS and the ANR.

References

1. Griffith, J. S. (1967) *Nature,* **215,** 1043.
2. Prusiner, S. B. (1982) *Science,* **216**, 136.
3. Cox, B. S. (1965) *Heredity,* **20**, 505.
4. Aigle, M., and Lacroute, F. (1975) *Mol. Gen. Genet.,* **136**, 327.
5. Derkatch., I. L., Bradley, M. E., Zhou, P., Chernoff, Y. O., and Liebman, S. W. (1997) *Genetics,* **147**, 507.
6. Coustou, V., Deleu, C., Saupe, S., and Begueret, J. (1997) *Proc. Natl. Acad. Sci. U.S.A.,* **94**, 9773.
7. Schoun, J., and Lacroute, F. (1969) *C. R. Acad. Sci.,* **269**,1412.
8. Cox, B. S., Tuite, M. F., and McLaughlin, C. S. (1988) *Yeast,* **4**, 159.
9. Wickner, R. B. (1994) *Science,* **264**, 566.
10. Hawthorne, D. C., and Mortimer, R. K. (1968) *Genetics,* **60**, 735.
11. Inge-Vechtomov, S. G., and Andrianova, V. M. (1970) *Genetika,* **6,** 103.
12. Frolova, L., Le Goff, X., Rasmussen, H. H., Chepereg, S., Drugeon, G., Kress, M., Arman, I., Haenni, I., Celis, A. L., Philippe, J. E., Jastesen, J., and Kisselev, L. (1994) *Nature,* **372**, 701.
13. Zhouravleva, G., Frolova, L., Le Goff, X., Le Guellec, R., Inge-Vechtomov, S., Kisselev, L., and Philippe, M. (1995) *EMBO J.,* **14**, 4065.

14. Stansfield, I., Jones, K. M., Kushnirov, V. V., Dagkesamanskaya, A. R., Poznyakovski, A. I., Paushkin, S. V., Nierras, C. R., Cox, B. S., Ter-Avanesyan, M. D., and Tuite, M. F. (1995) *EMBO J.,* **14,** 4365.

15. Patino, M. M., Liu, J. J., Glover, J. R., and Lindquist, S. (1996) *Science,* **273,** 622.

16. Kushnirov, V. V., Ter-Avanesyan, M. D., Telckov, M. V., Surguchov, A. P., Smirnov, V. N., and Inge-Vechtomov, S. G. (1988) *Gene,* **66,** 45.

17. Ter-Avanesyan, M. D., Kushnirov, V. V., Dagkesamanskaya, A. R., Didichenko, S. A., Chernoff, Y. O., Inge-Vechtomov, S. G., and Smirnov, V. N. (1993) *Mol. Microbiol.,* **7,** 683.

18. Ter-Avanesyan, M. D., Dagkesamanskaya, A. R., Kushnirov, V. V., and Smirnov, V. N. (1994) *Genetics,* **137,** 1339.

19. Derkatch, I. L., Chernoff, Y. O., Kushnirov, V. V., Inge-Vechtomov, S. G., and Liebman, S. W. (1996) *Genetics,* **144,** 1375.

20. Kong, C., Ito, K., Walsh, M. A., Wada, M., Liu, Y., Kumar, S., Barford, D., Nakamura, Y., and Song, H. (2004) *Mol. Cell,* **14,** 233.

21. Sondheimer, N., and Lindquist, S. (2000) *Mol. Cell,* **5,** 63.

22. Derkatch, I. L., Bradley, M. E., Hong, J. Y., and Liebman, S. W. (2001) *Cell,* **106,** 171.

23. Mitchell, A.P., and Magasanik, B. (1984) *Mol. Cell Biol.,* **4,** 2758.

24. Coschigano, P. M., and Magasanik, B. (1991) *Mol. Cell Biol.,* **11,** 822.

25. Courchesne, W. E., and Magasanik, B. (1988) *J. Bacteriol.,* **170,** 708.

26. Cooper T. G. (1982) in *The Molecular and Cellular Biology of the Yeast Saccharomyces: Metabolism and Gene Expression,* Vol 2 (ed. Strathern, J. N., Jones, E. W., and Broach J. R.), Cold Spring Harbor Laboratory, Cold Spring Harbor, NY, pp 39–99.

27. Magasanik B. (1992) in *The Molecular and Cellular Biology of the Yeast Saccharomyces cerevisiae,* Vol 2, 2nd edn (ed. Jones, E. W., Pringle, J. R., and Broach J. R.), Cold Spring Harbor Laboratory, Cold Spring Harbor, NY, pp 283–317.

28. Rai, R., Genbauffe, F., Lea, H. Z., and Cooper, T. G. (1987) *J. Bacteriol.,* **169,** 3521.

29. Lacroute, F. (1971) *J. Bacteriol.,* **106,** 519.

30. Thual, C., Komar, A. A., Bousset, L., Fernandez-Bellot, E., Cullin, C., and Melki, R. (1999) *J. Biol. Chem.,* **274,** 13666 .

31. Masison, D. C., and Wickner, R. B. (1995) *Science,* **270,** 93.

32. Perrett, S., Freeman, S. J., Butler, P. J., and Fersht, A. S. (1999) *J. Mol. Biol.,* **290,** 331.

33. Bousset, L., Belrhali, H., Janin, J., Melki, R., and Morera, S. (2001) *Structure,* **9**, 39.

34. Umland, T. C., Taylor, K. L., Rhee, S., Wickner, R. B., and Davies, D. R. (2001) *Proc. Natl. Acad. Sci. U.S.A.,* **98**, 1459.

35. Bousset, L., Belrhali, H., Melki, R., and Morera, S. (2001) *Biochemistry,* **40**, 13564.

36. Bai, M., Zhou, J. M., and Perrett, S. (2004) *J. Biol. Chem.,* **279**, 50025.

37. Glass, N. L., Jacobson, D. J., and Shiu, P. K. (2000) *Annu. Rev. Genet.,* **34**, 165.

38. Turcq, B., Deleu, C., Denayrolles, M., and Bégueret, J. (1991) *Mol. Gen. Genet.,* **228**, 3 .

39. Deleu, C., Clavé, C., and Bégueret, J. (1993) *Genetics,* **135**, 45.

40. Coustou, V., Deleu, C., Saupe, S. J., and Bégueret, J. (1999) *Genetics,* **153**, 1629 .

41. Bousset, L., Thomson, N. H., Radford, S. E., and Melki, R. (2002) *EMBO J.,* **21**, 2903.

42. Glover, J. R., Kowal, A. S., Schirmer, E. C., Patino, M. M., Liu, J. J., and Lindquist, S. (1997) *Cell,* **89**, 811 .

43. Krzewska, J., Tanaka, M., Burston, S. G., and Melki, R. (2007) *J. Biol. Chem.,* **282**, 1679.

44. Patel, B. K., and Liebman, S. W. (2007) *J. Mol. Biol.,* **365**, 773.

45. Dos Reis, S., Coulary-Salin, B., Forge, V., Lascu, I., Bégueret, J., and Saupe, S. J. (2002) *J. Biol. Chem.,* **277**, 5703 .

46. Thual, C., Bousset, L., Komar, A. A., Walter, S., Buchner, J., Cullin, C., and Melki, R. (2001) *Biochemistry,* **40**, 1764.

47. Serio, T. R., Cashikar, A. G., Kowal, A. S., Sawicki, G. J., Moslehi, J. J., Serpell, L., Arnsdorf, M. F., and Lindquist, S. L. (2000) *Science,* **289**, 1317.

48. Baxa, U., Cheng, N., Winkler, D. C., Chiu, T. K., Davies, D. R., Sharma, D., Inouye, H., Kirschner, D. A., Wickner, R. B., and Steven, A. C. (2005) *J. Struct. Biol.,* **150**, 170.

49. Astbury, W. T., and Dickinson, S. (1935) *Biochem. J.,* **29**, 2351.

50. Makin, O. S., and Serpell, L. C. (2005) *FEBS J.,* **272**, 5950.

51. Zurdo, J., Guijarro, J. I., and Dobson, C. M. (2001) *J. Am. Chem. Soc.,* **123**, 8141.

52. Sunde, M., Serpell, L. C., Bartlam, M., Fraser, P. E., Pepys, M. B., and Blake, C. C. (1997) *J. Mol. Biol.,* **273**, 729 .

53. Fandrich, M., and Dobson, C. M. (2002) *EMBO J.,* **21**, 5682.

54. Zhang, Z., and Smith, D. L. (1993) *Protein Sci.,* **2**, 522.

55. Johnson, R. J., Christodoulou, J., Dumoulin, M., Caddy, G. L., Alcocer, M. J., Murtagh, G. J., Kumita, J. R., Larsson, G., Robinson, C. V., Archer, D. B., Luisi, B., and Dobson, C. M. (2005) *J. Mol. Biol.,* **352**, 823.

56. Kheterpal, I., Chen, M., Cook, K. D., and Wetzel, R. (2006) *J. Mol. Biol.,* **361**, 785.

57. Kheterpal, I., Lashuel, H. A., Hartley, D. M., Walz, T., Lansbury, P. T, Jr., and Wetzel, R. (2003) *Biochemistry,* **42**, 14092 .

58. Wang, S. S., Tobler, S. A., Good, T. A., and Fernandez, E. J. (2003) *Biochemistry,* **42**, 9507 .

59. Del Mar, C., Greenbaum, E. A., L. Mayne, Englander, S. W., and Woods, V. L., Jr. (2005) *Proc. Natl. Acad. Sci. U.S.A.,* **102**, 15477 .

60. Nazabal, A., Dos Reis, S., Bonneu, M., Saupe, S. J., and Schmitter, J. M. (2003) *Biochemistry,* **42**, 8852.

61. Lu, X., Wintrode, P. L., and Surewicz, W. K. (2007) *Proc. Natl. Acad. Sci. U.S.A.,* **104**, 1510.

62. Bousset, L., Redeker, V., Decottignies, P., Dubois, S., Le Marechal, P., and Melki, R. (2004) *Biochemistry,* **43**, 5022.

63. Bousset, L., Briki, F., Doucet, J., and Melki, R. (2003) *J. Struct. Biol.,* **141**, 132 .

64. Ranson, N., Stromer, T., Bousset, L., Melki, R., and Serpell, L. C. (2006) *Protein Sci.,* **15**, 2481.

65. Redeker, V., Halgand, F., Le Caer, J. P., Bousset, L., Laprevote, O., and Melki, R. (2007) *J. Mol. Biol.,* **369**, 1113.

66. Gajdusek, D. C. (1988) *J. Neuroimmunol.,* **20**, 95.

67. Prusiner, S. B. (1991) *Science,* **252**, 1515.

68. Jarret, J. T., and Lansbury, P. T. (1993) *Cell,* **73**, 1055.

69. Telling, G. C., Scott, M., Mastrianni, J., Gabizon, R., Torchia, M., Cohen, F. E., DeArmond, S. J., and Prusiner, S. B. (1995) *Cell,* **83**, 79.

70. Uptain, S. M., Sawicki, G. J., Caughey, B., and Lindquist, S. (2001) *EMBO J.,* **20**, 6236.

71. Fernandez-Bellot, E., Guillemet, E., and Cullin, C. (2000) *EMBO J.,* **19**, 3215 .

72. Sawaya, M. R., Sambashivan, S., Nelson, R., Ivanova, M. I., Sievers, S. A., Apostol, M. I., Thompson, M. J., Balbirnie, M., Wiltzius, J. J., McFarlane, H. T., Madsen, A. Ø., Riekel, C., and Eisenberg, D. (2007) *Nature,* **447**, 453.

73. Nelson, R., Sawaya, M. R., Balbirnie, M., Madsen, A. Ø., Riekel, C., Grothe, R., and Eisenberg, D. (2005) *Nature,* **435**, 773.
74. Bu, Z., Shi, Y., Callaway, D. J., and Tycko, R. (2007) *Biophys. J.,* **92**, 594.
75. King, C. Y., and Diaz-Avalos, R. (2004) *Nature,* **428**, 319.
76. Tanaka, M., Chien, P., Naber, N., Cooke, R., and Weissman, J. S. (2004) *Nature,* **428**, 323.
77. Schlumpberger, M., Prusiner, S. B., and Herskowitz, I. (2001) *Mol. Cell. Biol.,* **21**, 7035.
78. Toyama, B. H., Kelly, M. J., Gross, J. D., and Weissman, J. S. (2007) *Nature,* **449**, 233.
79. Tessier, P. M., and Lindquist, S. (2007) *Nature,* **447**, 556.
80. Fay, N., Inoue, Y., Bousset, L., Taguchi, H., and Melki, R. (2003) *J. Biol. Chem.,* **278**, 30199.
81. Fay, N., Redeker, V., Savistchenko, J., Dubois, S., Bousset, L., and Melki, R. (2005) *J. Biol. Chem.,* **280**, 37149.
82. Fayard, B., Fay, N., David, G., Doucet, J., and Melki, R. (2006) *J. Mol. Biol.,* **356**, 843.
83. Lund, P. M., and Cox, B. S. (1981) *Genet. Res.,* **37**, 173.
84. Chernoff, Y. O., Lindquist, S. L., Ono, B., Inge-Vechtomov, S. G., and Liebman, S. W. (1995) *Science,* **268**, 880.
85. Moriyama, H., Edskes, H. K., and Wickner, R. B. (2000) *Mol. Cell. Biol.,* **20**, 8916.
86. Tuite, M. F., and Cox, B. S. (2003) *Nat. Rev. Mol. Cell Biol.,* **4**, 878.
87. Jones, G. W., and Masison, D. C. (2003) *Genetics,* **163**, 495.
88. Newnam, G. P., Wegrzyn, R. D., Lindquist, S. L., and Chernoff, Y. O. (1999) *Mol. Cell Biol.,* **19**, 1325.
89. Chernoff, Y. O., Newnam, G. P., Kumar, J., Allen, K., and Zink, A. D. (1999) *Mol. Cell Biol.,* **19**, 8103.
90. Roberts, B. T., Moriyama, H., and Wickner, R. B. (2004) *Yeast,* **21**, 107.
91. Loovers, H. M., Guinan, E., and Jones, G. W. (2007) *Genetics,* **175**, 621.
92. Sondheimer, N., Lopez, N., Craig, E. A., and Lindquist, S. (2001) *EMBO J.,* **20**, 2435.
93. Kushnirov, V. V., and Ter-Avanesyan, M. D. (1998) *Cell,* **94**, 13.
94. Glover, J. R., and Lindquist, S. (1998) *Cell,* **94**, 73.
95. Cox, B. S., Ness, F., and Tuite, M. F. (2003) *Genetics,* **165**, 23.
96. Shorter, J., and Lindquist, S. (2006) *Mol. Cell,* **23**, 425.

97. Krzewska, J., and Melki, R. (2006) *EMBO J.,* **25**, 822.
98. Savistchenko, J., Krzewska, J., Fay, N., and Melki, R. (2008) *J. Biol. Chem.,* **283**, 15732.
99. Speransky, V. V., Taylor, K. L., Edskes, H. K., Wickner, R. B., and Steven, A. C. (2001) *J. Cell Biol.,* **153,** 1327.
100. Tanaka, M., Chien, P., Naber, N., Cooke, R., and Weissman, J. S. (2004) *Nature,* **428**, 323.
101. King, C. Y., and Diaz-Avalos, R. (2004) *Nature,* **428**, 319.
102. Patel, B. K., and Liebman, S. W. (2007) *J. Mol. Biol.,* **365**, 773.
103. Brachmann, A., Baxa, U., and Wickner, R. B. (2005) *EMBO J.,* **24**, 3082.
104. Maddelein, M. L., Dos Reis, S., Duvezin-Caubet, S., Coulary-Salin, B., and Saupe, S. J. (2002) *Proc. Natl. Acad. Sci. U.S.A,* **99**, 7402.
105. Jiang, Y., Li, H., Zhu, L., Zhou, J. M., and Perrett, S. (2004) *J. Biol. Chem.,* **279**, 3361.
106. Nakayashiki, T., Ebihara, K., Bannai, H., and Nakamura , Y. (2001) *Mol. Cell,* **7**, 1121.
107. Kushnirov, V. V., Kochneva-Pervukhova, N. V., Chechenova, M. B., Frolova, N. S., and Ter-Avanesyan, M. D. (2000) *EMBO J.,* **19**, 324.
108. Chernoff, Y. O., Galkin, A. P., Lewitin, E., Chernova, T. A., Newnam, G. P., and Belenkiy, S. M. (2000) *Mol. Microbiol.,* **35**, 865.
109. Li, J. Y., Englund, E., Holton, J. L., Soulet, D., Hagell, P., Lees, A. J., Lashley, T., Quinn, N. P., Rehncrona, S., Björklund, A., Widner, H., Revesz, T., Lindvall, O., and Brundin, P. (2008) *Nat. Med.,* **14**, 501.
110. Kordower, J. H., Chu, Y., Hauser, R. A., Freeman, T. B., and Olanow, C. W. (2008) *Nat. Med.,* **14**, 504.

Chapter 7

Curli Extracellular Matrix

Izhack Cherny
Weizmann Institute of Science,
Rehovot 76100, Israel
izhack.cherny@weizmann.ac.il

Curli represent a class of highly aggregative thin surface fibres produced by many *Enterobacteriaceae*, serving as the proteinaceous component of the extracellular matrix. Curli are evident to confer multiple functions to bacteria, either in-host or non-host environments. Featuring a range of amyloidal characteristics, curli were classified as functional amyloids. This chapter introduces the curli extracellular matrix and their components, as well as the curli regulation and secretion. The chapter focuses on the amyloidogenic nature of curli, curli biogenesis, and the functions carried out by these fibres, in an attempt to shed light on how bacteria manage to exploit amyloid aggregation for their own benefit while avoiding the consequence associated with the uncontrolled formation of this state. Taken together, the importance of curli to aspects

The Functional Fold: Amyloid Structures in Nature
Edited by Suzi Jarvis and Anika Mostaert

ISBN 978-981-4267-40-3 (Hardcover), 978-981-4267-62-5 (eBook)
www.panstanford.com

of bacterial lifestyle is enormous — being a decisive factor in mediation of bacterial adhesion to a wide range of surfaces, host cells internalization, multi-cellular aggregation, pathogenesis, resistance to environmental stresses, and further to the overall biofilm architecture. It is remarkable that all these qualities are ascribed to a fold that only recently was considered one of nature's biggest evolutionary faults.

7.1 Bacterial Biofilms and the Curli Extracellular Matrix

Biofilm is defined as an organized surface-bound community of microorganisms sharing the same polymeric matrix.[1] Biofilms develop on virtually any material that comes in contact with any kind of naturally occurring moisture, from blood to salt water. Up to now, it has not been possible to devise a non-toxic coating method able to prevent biofilm formation. This is problematic when biofilm colonize biotic surfaces and on an assortment of devices in medical and food industry.

The transition from a planktonic to community lifestyle in a biofilm is ultimately made possible by the action of efficient organelles that act to initially adhere to the surface and later to maintain attachment between cells in their community. The curli fibres (also referred as Tafi, thin aggregative fimbriae, in *Salmonella* spp.) are defined as such proteinaceous adhesive surface organelles expressed by *Enterobacteriaceae* such as *Escherichia coli*, *Salmonella* spp., *Citrobacter* spp., and *Enterobacter sakazaki* implicated in surface colonization and biofilm formation. When visualized under light or electron microscopy, curli are observed as thin fimbriae, 3–12 nm in width and several micrometers in length. Curli represent only one component of the extracellular matrix. Other components are mostly cellulose fibres, but other surface structures may also be observed, such as type I fimbriae, flagella, and pili, which contribute to the biofilm development and purpose.

The course of biofilm formation is composed of several steps.[2,3] The first step is reversible surface attachment. Once irreversible adhesion has been established, micro-colonies are formed and eventually develop into a matured biofilm, characterized by distinct architecture that features mushroom-like structures, water channels, and pores. A mature biofilm allows bacteria dispersion to colonize new niches. The role of curli fibres in biofilm development is mostly in the initial adhesion steps and in the acquirement of the matured morphology.

Though formation of bacterial biofilms has been known for almost a century, it was only a decade ago that the curli extracellular matrix was recognized to be composed of amyloid fibres. Indeed, curli were among the first naturally occurring amyloids to be identified.[4] The mechanism of curli biogenesis and the molecular basis of its functions are further discussed in this chapter.

7.2 The Curli Gene Cluster

Curli formation requires the transcription of proteins from two divergent operons denoted as the *csg* gene cluster in *E. coli*, *csgBAC* and *csgDEFG* (Fig. 7.1). A homologous system that was identified in *Salmonella spp.*, comprised of the *agfBAC* and *agfDEFG* operons, was demonstrated to complement the *csg* genes.[5] The assembly of the fibres requires the secretion and extracellular self-assembly of the major curli subunit protein, CsgA (15 kDa), and involves a specific membrane bound nucleator protein, CsgB (15 kDa, the minor curli subunit). CsgB directs the polymerization of the curli fibres from the bacteria surface[6–8] and share sequence similarity with CsgA (~30% identity). The function of CsgC is yet unclear. However, it was suggested to facilitate AgfA (CsgA) correct extracellular assembly, as *agfC*$^{-}$ strain produced fimbriae with different morphology than that produced by the wild type.[9] The second operon, *csgDEFG*, entirely encodes for non-structural accessory proteins that are essential for curli biogenesis. CsgD acts as a positive regulator of *csgBAC* operon transcription.[6,10] CsgG is an outer membrane lipoprotein that mediates the secretion of CsgA and CsgB proteins.[4,11,12] CsgE and CsgF are periplasmic proteins that interact with CsgG in the outer membrane.[9,12] *E. coli* mutants that do not produce CsgE are defective in curli assembly and produce very low amount of CsgA fibres that are morphologically distinct from the wild-type fibres.[4,12] In contrast, *E. coli* bacteria that do not produce the CsgF protein are still able to secrete soluble CsgA and form fibres that are indistinguishable from wild-type fibres, but less abundant. It is speculated that *csgF* mutants are defective in CsgB functioning, while *csgE* encodes for a chaperon of CsgA.[4,12]

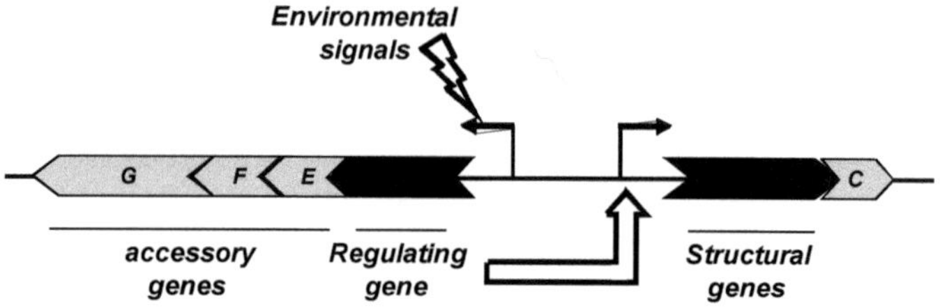

Figure 7.1 Schematic view of *csg* genes cluster. Correct secretion and assembly of curli extracellular matrix depends on the expression of both accessory and structural genes. A variety of environmental signals affect the transcription of curli accessory genes from the *csgDEFG* operon. Once induced, CsgD positively regulates the transcription of the structural genes from the second operon, *csgBAC*.

7.3 From Regulation to Secretion

7.3.1 Regulation of Curli Expression

Several environmental conditions were reported to promote curli expression: temperature below 30°C[13] (though it was shown that significant fraction of *E. coli* isolates from human blood cultures did produce curli at 37°C);[14] depletion in nutrients as phosphate, nitrogen, and iron;[15,16] oxygen tension;[5,15] presence of ethanol (4%);[15] and low osmolarity.[17,18] In contrast, excess of N-acetylglucosamine-6-phosphate (GlcNAc-6P) was reported to downregulate curli expression.[19] Several cellular factors were reported to influence curli expression: RpoS (stationary phase sigma factor) — induces curli expression during stationary phase; Crl — a transcription factor that is required for curli expression;[13] MlrA (MerR-like regulator) — a transcriptional regulator of *csgD* that positively regulates curli expression;[20] OmpR (outer membrane protein R) — together with EnvZ protein — they form a two component system that positively regulates *csgD* in response to changes in osmolarity;[16,21,22] CpxA/R two component system and Rcs — these systems negatively regulate curli expression in response to envelope stress;[17,23–27] IHF (integration host factor) this protein was shown to act as transcription factor of *csgD*;[28] HN-S (histone-like nucleoid structuring) — the function of this protein is still unclear as it was demonstrated both to positively[28] and negatively[29] regulate the expression of curli genes in *Salmonella typhimurium* and *E. coli*, respectively.

7.3.2 Stability and Secretion Machinery

Once expressed, the structural curli proteins (CsgA and CsgB) are directed to the periplasm. In the unpolymerized state, at least *in vitro*, these subunits were shown to adopt a predominant random coil (CsgA) or α-helical (AgfA) conformation rather than a β-rich one.[4,30] Furthermore, these molecules are secreted as soluble monomers.[4,7,9] The exact mechanism of their maintenance in this conformation is still unclear. It is expected that the aggregated form is toxic to the cell. The prevention of their aggregation might be achieved by (i) assigning a chaperon protein that would usher and/or accelerate CsgA and CsgB secretion through the outer membrane; (ii) assuring a short half-life of free subunits inside the cell. *csgCEFG* (*afgCEFG*) were suggested to possess a chaperon-like activity. AgfC was shown to be localized to the periplasm and be important for AgfA extracellular assembly.[9] A *csgE*$^-$ mutant produces significantly less abundant and somewhat structurally distinct fibrils.[4] In the absence of CsgG, a sharp decrease in the steady state levels of CsgA and CsgB in the periplasm was observed, indicating rapid degradation of the proteins.[11,12] Absence of CsgF results in aberrant or no secretion of CsgB.[4] Interestingly, overexpression of CsgG lipoprotein can compensate for CsgE deficiency (but not *vice versa*), suggesting that CsgE facilitates CsgG activity.

The *cgsEFG* accessory genes are essential for the proper secretion and biogenesis of curli. However, it is the CsgG lipoprotein that plays the principal role of secretion. CsgG forms oligomeric structures in the outer membrane, apparently forming pores in the membrane through which the CsgA and CsgB are transported to the cell surface.[12] CsgE and CsgF form a complex with CsgG at the outer membrane, and are considered to assist CsgG proper and efficient secretion.[12] The direction of CsgA to CsgG secretion machinery is solely dependent on a 22 amino acid sequence located at its N terminus. When assigned to a heterologous protein, this sequence can direct it to the cell surface as well.[12]

7.4 Curli Biogenesis: Mechanism, Kinetics, and Fibril Structure

7.4.1 Evidence to Being Naturally Occurring Amyloids

Curli fibres satisfy several mechanistic, structural, and tinctorial amyloid related features (Fig. 7.2). Curli fibres are insoluble in sodium dodecyl sulfate (SDS)[31] and resistant to protease digestion.[30] *In vivo* assembly depends on interaction with a nucleator protein (CsgB) to form amyloid-like fibres.[7] Similar to other amyloidogenic proteins, curli (and curli expressing bacteria) bind the dye Congo red (CR) and exhibit green birefringence as viewed by cross-polarization microscopy.[6] Curli were confirmed as naturally occurring amyloid fibres when they were shown to exhibit a red shift upon CR binding, to produce a significant fluorescence signal typical to amyloids upon binding to thioflavin T (ThT) dye, and to display spontaneous self-assembly by the purified monomers *in vitro* into amyloid-like fibrils. These fibres were substantially rich in β-sheet content according to electron microscopy and circular dichroism analysis.[4]

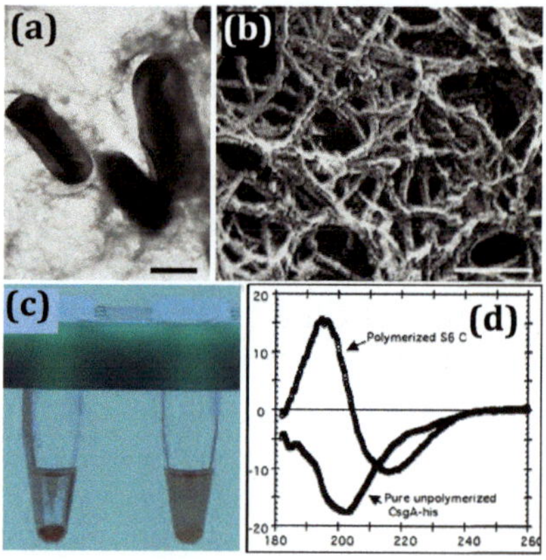

Figure 7.2 Amyloidal characteristics of curli. Transmission electron microscopy (TEM) micrograph (a) and of high-resolution deep-etch EM micrographs (b) of *E. coli* MC4100 expressing curli (bars: 500 nm and 60 nm, respectively). (c) Congo red dye binding by curliated (left) and non-curliated (right) *E. coli* MC4100. (d) CD spectrum of curli purified from MC4100 compared with that of soluble unpolymerized CsgA-his, indicating transition from random-coil to β-sheet rich structure. (b and d) Reprinted from Chapman *et al.*, *Science*, **295** (2005) with permission from AAAS. See also Colour Insert.

7.4.2 CsgA Self-Assembly: Insights from *in vitro and in silico* Research

Studies of the self-assembly of pure CsgA *in vitro* proved that it forms amyloids that share common features with "classic" amyloidogenic disease-associated proteins. CsgA was shown to display nucleation-dependent polymerization kinetics with lag, growth, and stationary phases.[32] As viewed by electron microscopy, fibre formation occurs already two hours after CsgA purification. The fibre formation process was demonstrated to be accelerated by the presence of preformed CsgA fibres significantly shortening the lag phase (process known as "seeding").[32]

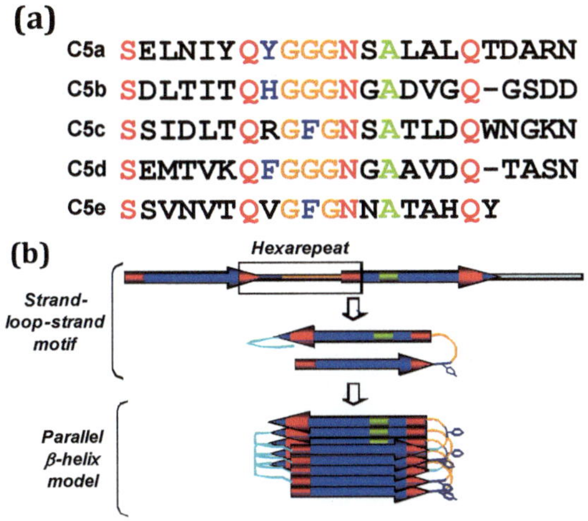

Figure 7.3 CsgA internal homology and structural model. (a) Internal alignment of the CsgA C-terminal domain. Conserved residues within repeating units (C5a–e) are coloured in red (polar), orange (glycines), blue (aromatics), and green (alanines). (b) Sequence corresponding representation of predicted stand-loop-strand motif. The assumed folding of the monomer into the parallel β-helix structure is illustrated. See also Colour Insert.

Interestingly, the structure of CsgA monomers after secretion was natively unfolded.[4,32] However, using A11 conformational antibodies that specifically recognize the transient species of amyloids[33] (such as Aβ and IAPP), transient intermediate-like structures could also be detected.[32] These were already present in the very beginning of the aggregation process (even after few minutes of incubation) in the form of monomers or dimers at the most, as determined by size exclusion chromatography. This finding led to two conclusions: (i) CsgA shares common polymerization intermediate with eukaryotic amyloids; (ii) CsgA has a strong predisposition toward formation of fibrillar structures, which apparently is encoded in its primary structure.

The innate tendency of CsgA to fibrillize is hypothesized to stem from the intramolecular repeats comprising its sequence, each acting as an independent amyloidogenic domain.[30,32,34] Five repeating units are included in the sequence of CsgA (and CsgB). These repeats, 19–23 amino acids in length, share 30% identity and contain four conserved polar residues of serine, glutamine, or asparagine every 5–7 amino acids (Fig. 7.3a). Computer homology modelling suggests that the repeats encode for five strand-loop-strand motifs that together fold into parallel β-helix structure of the monomer (Fig. 7.3b).[30,35] Accordingly, the structure is stabilized by a hydrophobic core and hydrogen bonds that occur between the conserved polar residues located at the tips of the turns in each repeating unit. Each turn region comprises conserved glycine residues to allow the necessary conformational flexibility, in addition to conserved aromatic residues (tyrosine, phenylalanine, or histidine) hypothesized to mediate interaction between turns regions.[34] In fact, the hexapeptide repeating sequences, composed of the turn sequence along with two flanking polar residues (glutamine or asparagine), were suggested to represent a minimal amyloidogenic motif, as they were shown to be sufficient to amyloid-like fibrillization.[34] The strand-loop-strand motif rather resembles the pair-of-sheets organization of the cross β-spine structure described for many disease-associated amyloids and fibril forming segments.[36,37] Wang *et al.*[32] suggested that the repeating units actually represent amyloidogenic domains, allowing efficient self-complementation and transition from soluble protein to amyloid fibre. This might explain the early recognition of intermediate-like conformations by A11 antibodies.[32]

7.4.3 *In Vivo* Curli Polymerization: Extracellular Nucleation–Precipitation Model

The process of curli biogenesis is believed to occur outside the bacterial cell through a proposed mechanism called extracellular nucleation–precipitation (ENP). According to the ENP hypothesis, both CsgA and CsgB subunits are secreted to the outer membrane as soluble proteins. While CsgA is released to the cellular milieu, CsgB is anchored to the membrane to provide CsgA subunits with the optimal nucleation point from the cell surface and to accelerate their precipitation into insoluble surface fibres. Though CsgA polymerization is feasible in the absence of CsgB nucleator, *in vivo* curli biogenesis is impractical with its absence. This model is supported by the following observations and demonstrations. (i) Interbacterial complementation of curli structural genes. A CsgA-deficient, CsgB-expressing cell (i.e., "receptor" cell) can direct the assembly of curli on its surface using CsgA monomers provided by a CsgB-deficient, CsgA-expressing cell (i.e., "donor" cell).[7] (ii) *In vitro* nucleation of CsgA by CsgB. Nevertheless, when secreted together with CsgA to the cellular milieu *in vivo* (defected in membrane localization), it displays diminished ability to nucleate CsgA into normal curli fibres.[8] (iii) Partial inhibition of curli formation by short peptides, rationally designed to block CsgA self-recognition motifs. These peptides were assumed to prevent CsgA monomers extracellular assembly into fibres.[34] (iv) Partial inhibition of curli formation in the presence of CsgA- or CsgB-specific antiserum.[9] Findings (iii) and (iv) provide firm evidence that curli structural subunits are secreted as soluble monomers.

7.5 Functional Role of Curli

Functional amyloids are unique since proteins are activated in their aggregated form. Before specifying curli functions, it may be appropriate to discuss functional amyloids and cell toxicity. Curli fibres, similar to other functional amyloids formed *in vivo*, are not known to display cell toxicity.[38] This may be explained by several features of curli. First, their unstructured conformation within the cells facilitates high turnover to avoid intracellular uncontrolled polymerization. Second, the structural monomers are hypothesized to preferentially adopt β-helix structures, allowing fast transition into its amyloidal fold. Third, curli assembly is highly regulated by a nucleator protein dedicated to localize and accelerate CsgA polymerization. Taken together, the presence of the assumed cytotoxic intermediates is significantly decreased.

Being amyloidogenic, curli proteins offer fast and efficient assembly as well as a degradation resistant coat. Curli are a critical component for the biofilm architecture, and further promote multi-cellular aggregation and pellicle formation, by mediating cell–cell interaction within a colony, which is important to produce the characteristic thick biofilm.[22,39–42]

The curli matrix layer confers the bacteria with protection and enhanced survival outside on the host one hand, and persistence and virulence tools within the host on the other hand. Indeed, *Enterobacteriaceae* may spend long periods of their life cycle outside animal or plant hosts, where they may encounter suboptimal and stressful conditions. One way to deal with these changes is to attach to solid surfaces in an aqueous environment and form biofilm, in this way maximizing metabolic breakdown and stress management.[43] Accordingly, curli amyloids were shown to assist bacteria to attach to abiotic inert surfaces, including non-hydrophobic ones.[44,45] Among the amyloid-facilitated binding surfaces are glass, stainless steel, polystyrene, and Teflon.[22,45,46] Furthermore, curli expression was demonstrated to be protective against a range of antimicrobial agents, such as hydrogen peroxidase, quaternary ammonium, chlorine, and to grant enhanced survival during desiccation in the absence of nutrients.[46-48] This facilitates bacterial colonization and persistence in medical and technical devises, leading to infections and biodeterioration. For example, *E. coli* is the most common bacterium found in biofilms that have developed on catheters introduced into the urinary tract.[49]

Within the host, where nutritional resources are readily available, bacteria use biofilm to persist and further disseminate throughout the host. In this environment, either animal or plant, curli fibres function to promote stable adhesion to various tissues.[50–52] Hence, curli have a major role in bacterial pathogenicity. A list of evidence supports the importance of curli during the infectious process. Curli are expressed at 37°C in biofilms of many clinical isolates.[41] Among clinical isolates, curli are expressed by most enterohaemorrhagic, enterotoxigenic, and sepsis strains, whereas enteroinvasive and enteropathogenic strains do not express curli.[18,53] Curli was demonstrated to interact with a variety of host proteins, many of which are proposed to facilitate bacterial dissemination through the host. For instance, curli binding to extracellular matrix and contact-phase proteins (such as laminin, fibrinogen, bradykinin, and factor XII) was suggested to deplete coagulation factors.[6,53–55] In addition, curli are able to activate plasminogen by capturing it simultaneously with tissue plasminogen activator, leading to the proximate degradation of soft tissues.[56] Furthermore, curli interact with molecules of the immune system, such as the Toll-like receptors and major histocompatibility complex class I. The resulted activation of the host innate immune system, as well as release of bradykinin, stimulate the inflammatory response.[14,54,57–59] Finally, several findings notably indicated the role of curli in the infection process: curli expressing *E. coli* and *S. typhimurium* strains, either pathogenic or artificially expressing, display better adherence and invasion into epithelial cells;[41,50,51,59–61] inhibition of curli formation by rationally designed peptides correlated with lower internalization rates;[34] and presence of curliated *E. coli* was related to sudden infant death syndrome.[62] Taken together, curli properties act to increase the spread of bacteria to surrounding tissues, promote infections, and contribute to the symptoms of sepsis.

The mechanism by which curli mediate invasion was proposed to involve serum proteins binding, primarily fibronectin. When inert particles such as latex beads are enveloped with fibronectin, they can be taken up by cells.[63] Curli fibres are assumed to take advantage of this process, since a peptide containing the arginine-glycine-aspartate (RGD) motif, responsible for interaction of fibronectin with cellular integrins, was shown to strongly inhibit curli-mediated internalization.[64] In an attempt to determine the regions within CsgA sequence that mediate binding to fibronectin, two peptides corresponding to distant regions in the primary sequence were identified, which were further shown to activate the pro-inflammatory contact system.[65] This substantiates that curli functionality does not merely stem from the amyloid fold.

7.6 Summary

Curli are an extraordinary example of how living microorganisms harness the functional advantages offered by the amyloid fold, while utterly dominating its formation and limiting its cellular toxicity. These functional amyloids also demonstrate how specific properties (as selective tissue binding or nucleation rather than polymerization propensity) could be integrated into amyloidogenic sequences.

Indeed, the value of amyloids as an extracellular component is significant in view of a recent finding showing that amyloids are widely distributed among natural biofilm, including bacteria from several phyla, such as *Proteobacteria, Bacteriodetes, Firmicutes, Chloroflexi*, and *Actinobacteria*.[44]

The discovery of curli amyloid may have important implications. First, curli may serve as an exceptional model for the study of amyloid formation and disease-associated amyloidogenesis, including the development of new strategies for the treatment of amyloid diseases. Second, activation of enzymatic processes in mammalian blood and ECM holds a promise for the development of new therapeutic agents. Finally, from biotechnological point of view, curli are a fascinating biomaterial. Beyond the mechanical and functional qualities, curli assembly is directed, highly regulated, yet efficient and may be easily produced on a large scale. Such characteristics are much desired for prospective nanobiotechnological applications.

Acknowledgement

I thank Dr. Jermont Chen, Dr. Shona C. Patel, and Dr. Sharon Gilead for valuable comments on the early drafts.

References

1. Costerton, J. W., Lewandowski, Z., Caldwell, D. E., Korber, D. R., and Lappin-Scott, H. M. (1995) *Annu. Rev. Microbiol.*, **49**, 711.
2. Van Houdt R., and Michiels, C. W. (2005) *Res. Microbiol.*, **156**, 626.
3. Stoodley, P., Sauer, K., Davies, D. G., and Costerton, J. W. (2002) *Annu. Rev. Microbiol.*, **56**, 187.
4. Chapman, M. R., Robinson, L. S., Pinkner, J. S., Roth, R., Heuser, J., Hammar, M., Normark, S., and Hultgren, S. J. (2002) *Science*, **295**, 851.
5. Romling, U., Bian, Z., Hammar, M., Sierralta, W. D., and Normark, S. (1998) *J. Bacteriol.*, **180**, 722.
6. Hammar, M., Arnqvist, A., Bian, Z., Olsen, A., and Normark, S. (1995) *Mol. Microbiol.*, **18**, 661.
7. Hammar, M., Bian, Z., and Normark, S. (1996) *Proc. Natl. Acad. Sci. USA*, **93**, 6562.
8. Hammer, N. D., Schmidt, J. C., and Chapman, M. R. (2007) *Proc. Natl. Acad. Sci. USA*, **104**, 12494.

9. Gibson, D. L., White, A. P., Rajotte, C. M., and Kay, W. W. (2007) *Microbiology*, **153**, 1131.
10. Gerstel U., and Romling, U. (2003) *Res. Microbiol.*, **154**, 659.
11. Loferer, H., Hammar, M., and Normark, S. (1997) *Mol. Microbiol.*, **26**, 11.
12. Robinson, L. S., Ashman, E. M., Hultgren, S. J., and Chapman, M. R. (2006) *Mol. Microbiol.*, **59**, 870.
13. Arnqvist, A., Olsen, A., Pfeifer, J., Russell, D. G., and Normark, S. (1992) *Mol. Microbiol.*, **6**, 2443.
14. Bian, Z., Brauner, A., Li, Y., and Normark, S. (2000) *J. Infect. Dis.*, **181**, 602.
15. Gerstel U., and Romling, U. (2001) *Environ. Microbiol.*, **3**, 638.
16. Romling, U., Sierralta, W. D., Eriksson, K., and Normark, S. (1998) *Mol. Microbiol.*, **28**, 249.
17. Prigent-Combaret, C., Brombacher, E., Vidal, O., Ambert, A., Lejeune, P., Landini, P., and Dorel, C. (2001) *J. Bacteriol.*, **183**, 7213.
18. Olsen, A., Arnqvist, A., Hammar, M., Sukupolvi, S., and Normark, S. (1993) *Mol. Microbiol.*, **7**, 523.
19. Barnhart, M. M., Lynem, J., and Chapman, M. R. (2006) *J. Bacteriol.*, **188**, 5212.
20. Brown, P. K., Dozois, C. M., Nickerson, C. A., Zuppardo, A., Terlonge, J., and Curtiss, R. 3rd. (2001) *Mol. Microbiol.*, **41**, 349.
21. Pratt, L. A., Hsing, W., Gibson, K. E., and Silhavy, T. J. (1996) *Mol. Microbiol.*, **20**, 911.
22. Vidal, O., Longin, R., Prigent-Combaret, C., Dorel, C., Hooreman, M., and Lejeune, P. (1998) *J. Bacteriol.*, **180**, 2442.
23. Raivio T. L., and Silhavy, T. J. (2001) *Annu. Rev. Microbiol.*, **55**, 591.
24. Hung, D. L., Raivio, T. L., Jones, C. H., Silhavy, T. J., and Hultgren, S. J. (2001) *EMBO J.*, **20**, 1508.
25. Ferrieres L., and Clarke, D. J. (2003) *Mol. Microbiol.*, **50**, 1665.
26. Jubelin, G., Vianney, A., Beloin, C., Ghigo, J. M., Lazzaroni, J. C., Lejeune, P., and Dorel, C. (2005) *J. Bacteriol.*, **187**, 2038.
27. Vianney, A., Jubelin, G., Renault, S., Dorel, C., Lejeune, P., and Lazzaroni, J. C. (2005) *Microbiology*, **151**, 2487.
28. Gerstel, U., Park, C., and Romling, U. (2003) *Mol. Microbiol.*, **49**, 639.
29. Arnqvist, A., Olsen, A., and Normark, S. (1994) *Mol. Microbiol.*, **13**, 1021.
30. Collinson, S. K., Parker, J. M., Hodges, R. S., and Kay, W. W. (1999) *J. Mol. Biol.*, **290**, 741.

31. Collinson, S. K., Emody, L., Muller, K. H., Trust, T. J., and Kay, W. W. (1991) *J. Bacteriol.*, **173**, 4773.
32. Wang, X., Smith, D. R., Jones, J. W., and Chapman, M. R. (2007) *J. Biol. Chem.*, **282**, 3713.
33. Kayed, R., Head, E., Thompson, J. L., McIntire, T. M., Milton, S. C., Cotman, C. W., and Glabe, C. G. (2003) *Science*, **300**, 486.
34. Cherny, I., Rockah, L., Levy-Nissenbaum, O., Gophna, U., Ron, E. Z., and Gazit, E. (2005) *J. Mol. Biol.*, **352**, 245.
35. White, A. P., Collinson, S. K., Banser, P. A., Gibson, D. L., Paetzel, M., Strynadka, N. C., and Kay, W. W. *J.* (2001) *Mol. Biol.*, **311**, 735.
36. Sawaya, M. R., Sambashivan, S., Nelson, R., Ivanova, M. I., Sievers, S. A., Apostol, M. I., Thompson, M. J., Balbirnie, M., Wiltzius, J. J., McFarlane, H. T., Madsen, A. O., Riekel, C., and Eisenberg, D. (2007) *Nature*, **447**, 453.
37. Nelson R., and Eisenberg, D. (2006) *Adv. Protein. Chem.*, **73**, 235.
38. Fowler, D. M., Koulov, A. V., Balch, W. E., and Kelly, J. W. (2007) *Trends. Biochem. Sci.*, **32**, 217.
39. Austin, J. W., Sanders, G., Kay, W. W., and Collinson, S. K. (1998) *FEMS Microbiol. Lett.*, **162**, 295.
40. Romling, U., Rohde, M., Olsen, A., Normark, S., and Reinkoster, J. (2000) *Mol. Microbiol.*, **36**, 10.
41. Kikuchi, T., Mizunoe, Y., Takade, A., Naito, S., and Yoshida, S. (2005) *Microbiol. Immunol.*, **49**, 875.
42. Jonas, K., Tomenius, H., Kader, A., Normark, S., Romling, U., Belova, L. M., and Melefors, O. (2007) *BMC Microbiol.*, **7**, 70.
43. Otto K., and Hermansson, M. (2004) *J. Bacteriol.*, **186**, 226.
44. Larsen, P., Nielsen, J. L., Dueholm, M. S., Wetzel, R., Otzen, D., and Nielsen, P. H. (2007) *Environ. Microbiol.*, **9**, 3077.
45. Pawar, D. M., Rossman, M. L., and Chen, J. (2005) *J. Appl. Microbiol.*, **99**, 418.
46. Uhlich, G. A., Cooke, P. H., and Solomon, E. B. (2006) *Appl. Environ. Microbiol.*, **72**, 2564.
47. Ryu J. H., and Beuchat, L. R. (2005) *Appl. Environ. Microbiol.*, **71**, 247.
48. White, A. P., Gibson, D. L., Kim, W., Kay, W. W., and Surette, M. G. (2006) *J. Bacteriol.*, **188**, 3219.
49. Hessen M. T., and Kaye, D. (1989) American Society for Microbiology, Washington, D.C., p.199.
50. Sukupolvi, S., Lorenz, R. G., Gordon, J. I., Bian, Z., Pfeifer, J. D., Normark, S. J., and Rhen, M. (1997) *Infect. Immun.*, **65**, 5320.
51. Kim S. H., and Kim, Y. H. (2004) *J. Vet. Sci.*, **5**, 119.

52. Barak, J. D., Gorski, L., Naraghi-Arani, P., and Charkowski, A. O. (2005) *Appl. Environ. Microbiol.*, **71**, 5685.

53. Ben Nasr, A., Olsen, A., Sjobring, U., Muller-Esterl, W., and Bjorck, L. (1996) *Mol. Microbiol.*, **20**, 927.

54. Herwald, H., Morgelin, M., Olsen, A., Rhen, M., Dahlback, B., Muller-Esterl, W., and Bjorck, L. (1998) *Nat. Med.*, **4**, 298.

55. Olsen, A., Arnqvist, A., Hammar, M., and Normark, S. (1993) *Infect. Agents Dis.*, **2**, 272.

56. Sjobring, U., Pohl, G., and Olsen, A. (1994) *Mol. Microbiol.*, **14**, 443.

57. Tukel, C., Raffatellu, M., Humphries, A. D., Wilson, R. P., Andrews-Polymenis, H. L., Gull, T., Figueiredo, J. F., Wong, M. H., Michelsen, K. S., Akcelik, M., Adams, L. G., and Baumler, A. J. (2005) *Mol. Microbiol.*, **58**, 289.

58. Olsen, A., Wick, M. J., Morgelin, M., and Bjorck, L. (1998) *Infect. Immun.*, **66**, 944.

59. Wang, X., Rochon, M., Lamprokostopoulou, A., Lunsdorf, H., Nimtz, M., and Romling, U. (2006) *Cell. Mol. Life. Sci.*, **63**, 2352.

60. Gophna, U., Barlev, M., Seijffers, R., Oelschlager, T. A., Hacker, J., and Ron, E. Z. (2001) *Infect. Immun.*, **69**, 2659.

61. Uhlich, G. A., Keen, J. E., and Elder, R. O. (2002) *Infect. Immun.*, **70**, 395.

62. Goldwater P. N., and Bettelheim, K. A. (2002) *J. Med. Microbiol.*, **51**, 1009.

63. Ozeri, V., Rosenshine, I., Mosher, D. F., Fassler, R., and Hanski, E. (1998) *Mol. Microbiol.*, **30**, 625.

64. Gophna, U., Oelschlaeger, T. A., Hacker, J., and Ron, E. Z. (2002) *FEMS Microbiol. Lett.*, **212**, 55.

65. Olsen, A., Herwald, H., Wikstrom, M., Persson, K., Mattsson, E., and Bjorck, L. (2002) *J. Biol. Chem.*, **277**, 34568.

Chapter 8

Amyloid-Based Natural Adhesives and Cements

Anika S. Mostaert[a,b] and Suzanne P. Jarvis[b,c]
[a]*School of Biology and Environmental Science,*
[b]*Conway Institute of Biomolecular and Biomedical Research, and*
[c]*School of Physics, University College Dublin, Dublin 4, Ireland*
anika.mostaert@ucd.ie; suzi.jarvis@ucd.ie

Functional amyloid has recently emerged as an underlying strategy for providing mechanical strength to many natural adhesives and cements from both prokaryotic and eukaryotic organisms. Specifically, functional amyloid has been identified in adhesives extruded from marine invertebrate, algae, and as a component of barnacle cement. In addition, cursory evidence exists to suggest that a number of other adhesive systems may also utilize functional amyloid. Atomic force microscopy has allowed nanoscale mechanical responses of functional amyloid fibrils to be made directly from the natural adhesive matrix, and has led to unexpected insights into the function of amyloid in natural materials. The molecular level origin of cohesive strength was found to be associated with the

The Functional Fold: Amyloid Structures in Nature
Edited by Suzi Jarvis and Anika Mostaert

ISBN 978-981-4267-40-3 (Hardcover), 978-981-4267-62-5 (eBook)
www.panstanford.com

generic amyloid intermolecular β-sheet structure, while adhesive strength was related to adhesive residues external to the amyloid core. The mechanistic link of amyloid-based cohesive and adhesive strength is now expected to be more widespread amongst natural adhesives than previously thought. These remarkably robust and highly ordered protein assemblies in biological adhesives provide inspiration for the biomimetic development of a new generation of adhesives for use in medicine, biotechnology, and a range of other applications.

8.1 Introduction to Natural Adhesives

Nature has evolved extraordinary adhesive mechanisms to promote the attachment of a wide diversity of organisms to many materials. These strategies are essential for successful completion of an organism's life cycle, thereby ensuring its survival in aquatic or terrestrial habitats. Interest in natural adhesives has primarily focused on organisms from aquatic environments, particularly marine, to develop anti-fouling strategies, or to develop aqueous-based biomimetic adhesives.[1,2] These marine adhesives are remarkably effective in attaching to virtually any damp, wet, or submerged surface, with strong, flexible bonds. Sessile marine organisms depend upon tenacious permanent adhesion (e.g., barnacles, mussels, and algae) to deal with shear forces in the ocean, or at the tidal interface, while motile organisms rely instead on strong, but transitory, adhesion to different surfaces (e.g., echinoderms, gastropods, gliding diatoms) to readily enable locomotion.[3] Whether temporary or permanent, nature provides many examples of successful attachment to various materials in a moist environment[2] (Fig. 8.1).

Due to the chemical and structural complexity of biological adhesives and cements, precise mechanisms have been clarified for only a few systems. For example, the strong, permanent attachment of the marine mussel *Mytilus edulis* has been found to be reliant upon an adhesive with a high content of the modified amino acid 3,4-dihydroxy-L-phenylalanine,[4,5] whereas barnacles adhere initially in cyprid larval form via o-quinone cross-linking,[6] followed by an adult cement comprising three groups of proteins that contain high levels of serine, threonine, glycine, and alanine.[6,7]

Figure 8.1 Sub-aerial microalgae colonizing natural and anthropogenic surfaces. Green microalgae (Chlorophyta) can rapidly grow as extensive biofilms on natural surfaces such as (a) on this tree in Clontarf, Dublin, Ireland, (b) on a range of artificial surfaces such as shown here as a conspicuous biofilm on a wooden fence in Rostock, Germany, and (c) on concrete surfaces of a staircase at Howth, Dublin, Ireland. The way they attach so strongly (even in inhospitable environmental conditions) has been attributed to amyloid structures in their adhesive matrix. See also Colour Insert.

Despite their great potential for inspiring biomimetic applications, there are a number of reasons why many biological adhesives are only recently becoming better understood. Although some organisms secrete a relatively large quantity of extracellular polymeric substances to adhere, there are many smaller attachment organisms (e.g., protists, fungi, and microalgae) which only secrete miniscule amounts of complex adhesive. Thus only very small amounts are available which can be isolated for biochemical analyses. This poses an enormous challenge for the identification and characterization of specific adhesive polymers, and further difficulties arise due to their high insolubility and rapid curing. Overcoming these challenges signals the turning point for extracting the exact working mechanisms and developing a full understanding of biological adhesives and cements. Significant progress can be made by approaching the field in an interdisciplinary way, combining a knowledge of ecology to identify potentially useful model organisms with biochemistry expertise to analyse the adhesives and cements, and mechanics to interpret their structure and material properties.

By using modern methods such as the tools of nanotechnology for studying the molecular mechanisms of adhesives, we have an opportunity to build upon the basic biochemical and mechanical principles involved in adhesion.[8] As research into natural adhesives and cements progresses, similarities in underlying principles have begun to emerge, while concurrently more variations in the details becomes clear. This has provided some surprising insights into an unexpected commonality amongst many adhesives and cements, as well as emphasizing the multitude of ways that adhesion can be achieved. This chapter focuses on the identification and mechanical properties of one such structural commonality between otherwise unconnected organisms: *Functional amyloid.*

8.2 Amyloid in Natural Adhesives

The previous chapter (Chapter 7) describes the detection of amyloid amongst prokaryotes, where *Escherichia coli* was shown to produce biofilm-associated amyloid fibrils known as curli.[9] A specific functionality was not identified for curli at that time. Curli was described as a surface protein polymer that mediated interactions important for biofilm formation, host cell colonization, adhesion, and cell aggregation.[9,10] Several other bacteria (e.g., *Salmonella*) have since been found to produce other families of amyloid fibrils (see review of Otzen and Nielsen, 2007),[11] and it is believed that multiple roles exist for amyloid as its widespread occurrence was discovered in further bacterial species from different habitats.[12,13]

Functional amyloid was not identified in natural adhesives until 2006 when a small, green algae, *Prasiola linearis*, was shown to produce amyloid in the adhesive matrix associated with the holdfast region that attaches permanently to surfaces in coastal lagoons and saltmarshes.[14] This was the first report of an amyloid-based biological adhesive and a specific mechanical functionality was associated with the amyloid component.[14] A mechanical response was proposed relative to the molecular structure of the generic amyloid fibril form found in natural adhesives using atomic force microscopy (AFM). The following year, amyloid was found in the adhesive of another algal species, the filamentous terrestrial green algae *Klebsormidium flaccidum*.[15] Additional characteristics of benefit to the mechanical properties of the adhesive were revealed, leading to the suggestion that amyloid fibrils may provide a generic mechanism for mechanical strength in natural adhesives.

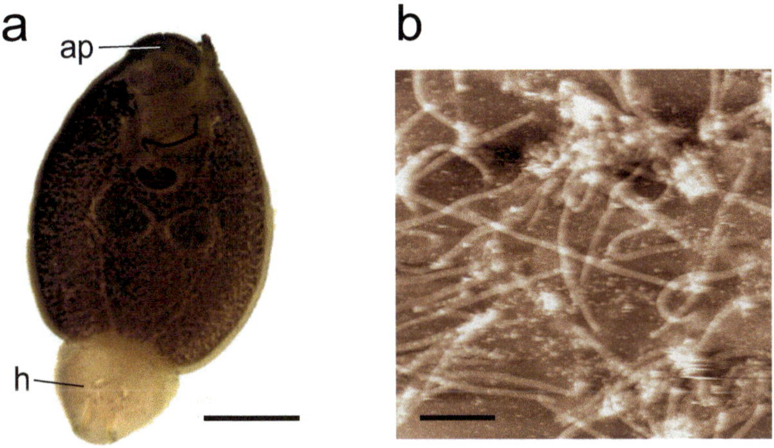

Figure 8.2 (a) Light micrograph of the marine parasite *Entobdella soleae* (ventral view). Anterior pads (ap) bind to fish skin using a temporary adhesive secretion, and the haptor (h) attaches to host skin by suction (scale bar, 1 mm). (b) Atomic force microscopy (AFM) image of fibres obtained directly from an adhesive pad print of *E. soleae* without purification or additional isolation (scale bar, 3 μm). The dark to light height range = 0–15 nm. From A. S. Mostaert, R. Crockett, G. Kearn, I. Cherny, E. Gazit, L. C. Serpell, and S. P. Jarvis, Mechanically functional amyloid fibrils in the adhesive of a marine invertebrate as revealed by Raman spectroscopy and atomic force microscopy, *Archives of Histology and Cytology*, **72**, 199–207 (2009), reprinted by permission of the publisher (International Society of Histology and Cytology).[17] See also Colour Insert.

Since then, two species of unicellular green algae (*Cocccomyxa* sp. and *Glaphyrella trebouxiodes*) growing in damp, sub-aerial habitats in terrestrial environments (Fig. 8.1), have also been shown to produce amyloid in their permanently attached adhesive pads beneath each individual algal cell.[16] These amyloid structures were studied in detail with AFM to characterize their nanomechanical properties, and link this to their possible function (see also Section 8.3). In the same year, an amyloid-based adhesive was found to be secreted from the parasitic marine flatworm, *Entobdella soleae*[17] that attaches to the skin of the common sole (*Soleae solea*). This parasite produces a temporary adhesive that appears to form a (sub)-monolayer coverage of proteinaceous structures without any additional matrix for attachment (Fig. 8.2).[18,19] This adhesive film, of only 10 nm thickness, supports a parasite up to 5 mm in length which will be subjected to significant shear forces while attached to the host fish in the ocean. This ratio of supported volume to adhesive volume is extraordinary. As described later in this chapter, the functional amyloid from the adhesives of the algae and parasite are surprisingly similar in terms of their mechanical properties.

While the discovery of amyloid-based adhesives was initially inferred from AFM mechanical responses attributed to the repetitive protein elements,[14,15] the amyloid-specific dyes thioflavin-T and Congo red were used as further evidence whenever possible.[14,16,17] Detection of amyloid is commonly based on the binding of these dyes to the β-sheets of the protein,[20] and both could be used directly on the adhesives *in situ*. It should be noted that neither is perfect with regards to its specificity and usability. However, both are good support for the presence of amyloid when used in combination. Raman spectroscopy was used to provide additional evidence that the structures characterized by AFM were amyloid,[16,17] and could also be performed on an adhesive matrix *in situ*.

Definitive verification of the presence of amyloid is a major issue particularly with secreted materials of complex composition. Instead of a single method being used it is often only possible to build up evidence based on a range of complementary techniques (such as those detailed above) and when possible the additional application of techniques, such as Fourier transform infrared (FTIR) spectroscopy and far-UV circular dichroism (CD) spectroscopy. In some cases, treatment with extremely strong organic acids may release monomers that can be further purified for SDS-PAGE and sequenced by mass spectrometry.[11] This, however, requires pure cultures, ample expression of amyloid, and an efficient protocol optimized for an individual organism, due to the robust and insoluble structures involved.[11] For future studies, there exists a challenging need to refine and develop procedures to identify amyloid more clearly and readily in natural materials, such as biological adhesives and cements.

In view of these significant difficulties, the potential presence of amyloid in adhesives and cements in former studies may have been overlooked simply because the detection methods were unsuitable, difficult, or unavailable. This may have been the case for barnacle cement. Barnacle cement is permanent, extremely insoluble, and composed of proteinaceous nanofibrils.[1] It was noted as having some similar characteristics to an amyloid plaque by Kamino *et al.*,[7] but not confirmed as being amylogenic until a broad range of analytical tools were later employed.[21,22] Using energy dispersive X-ray analysis and FTIR spectroscopy, in addition to the chemical staining and AFM, Sullan *et al.*[21] revealed amyloid in cement from the barnacle *Balanus amphitrite* (*Amphibalanus amphitrite*), and described that the amyloid component provided toughness to the cement. Barlow *et al.*[22] used similar analytical techniques, and included CD spectroscopy, to characterize the primary and secondary cement of the same barnacle species, finding both cements to be composed largely of amyloid fibres along with an unidentified globular protein component.

These recent reports of amyloid-based adhesives and cements indicate that they are likely to be widespread in nature, in different environments, and produced by phylogenetically different organisms.

8.3 AFM as a Tool for Exploring Natural Adhesives

One approach to investigating the structure–function relationships of individual biomolecules of natural adhesives is to use a local force probe, known as AFM.[23] AFM measures intermolecular forces between a sharp tip located at the end of a cantilever, and the sample surface of interest, and builds up a three-dimensional image corresponding to the surface topography. The capabilities of AFM instrumentation facilitate molecular resolution imaging under physiological conditions with minimal to no sample preparation.[24] In other words, single adhesive polymers may be imaged *in situ*, without any fixation or subsequent processing.

This ability of AFM to directly probe small volumes of biological material with nanoscale precision, in an ambient environment, makes it ideally suited to exploring natural adhesives and cements. Through single molecule force spectroscopy, AFM also allows nanoscale characterization of the mechanical properties of biological materials such as their adhesive strength and elasticity.[24] For many biological molecules, response to an externally applied force is important functionally and structurally and thus force spectroscopy has become a valuable analytical tool for the investigation of physical and mechanical properties of single biomolecules in aqueous environments.[25,26]

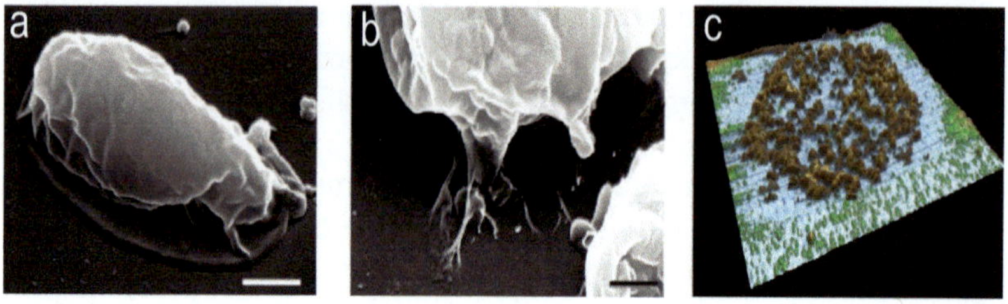

Figure 8.3 Images of algal adhesives. (a and b) Scanning electron microscope images of single microalgal cells attached to glass coverslips. A layer of extracellular polymeric substance (EPS) covers the surface of each cell, and differentiated region forming the adhesive pad can be seen beneath the cells. Scale bar = 1 μm. (c) Atomic force microscope image of the adhesive beneath a single microalgal cell after the cell was removed from the coverslip surface. Scan size = 5 μm. (a and b) from A. S. Mostaert, C. Giordani, R. Crockett, U. Karsten, R. Schumann, and S. P. Jarvis, Characterisation of amyloid nanostructures in the natural adhesive of unicellular subaerial algae, *Journal of Adhesion*, **85**, 465 (2009), reprinted by permission of the publisher (Taylor & Francis Ltd, http://www.tandf.co.uk/journals).[16] See also Colour Insert.

AFM was initially found to be particularly useful for exploring the mechanical properties of natural adhesives from marine organisms, as it allowed measurements to be made on materials in their hydrated, native state under aqueous conditions, without isolation or fixation. Examples include those of Dammer *et al.*[27] for cell-adhesion proteoglycans of a marine sponge, Smith *et al.*[28] for the lustrin A adhesive binding abalone nacre, and Callow *et al.*[29] for the green alga, *Ulva linza* (syn. *Enteromorpha linza*), a common marine fouling organism. Wetherbee and co-workers later worked on the nanomechanical properties of adhesives from motile and sessile diatoms[30–34] and Jarvis and co-workers focused on adhesives from sub-aerial algae (Fig. 8.3)[14–16] and a marine parasite.[17] In the latter group, the unusually repetitive mechanical response of the adhesive led the researchers into the new area of characterizing mechanically functional amyloid in natural adhesives.

Using AFM the elastic and inelastic response of amyloid fibrils may be measured and related to the response of monomers within the amyloid core. The quantitative variation in mechanical response at the molecular level has been measured to characterize amyloid in biological adhesives, that is, functional amyloid formed *ex vivo*.[17] A detailed mechanical understanding of the varying amyloid structures that underpin a broad range of natural adhesives has been developed.[15,17]

8.3.1 Nanomechanical Properties of Natural Adhesives: Amyloid Structures Provide Mechanical Strength and Toughness

One of the most important evaluation criteria of materials is their mechanical stability to applied load and ultimately how they fail or break as the load increases. To evaluate this for amyloid based adhesives, AFM has been used to gently pull the material apart. Locations where amyloid structures may occur are highlighted due to the very repetitive nature of the mechanical response (Fig. 8.4) and sometimes also from AFM images of the material which can indicate the presence of fibrillar structures (Fig. 8.2b). Pulling apart any repetitive structure is clearly indicated in force–extension curves produced when an AFM tip binds (non-specific adsorption) to the sample surface, and the substrate and tip are moved apart under an increasing tensile load, thus unravelling the structure until it either breaks or detaches from the tip or substrate[15] (Fig. 8.4b).

Resultant force–distance curves, show a repetitive mechanical response corresponding to the unravelling of the repetitive structural units of amyloid on retraction of the AFM tip. Observed as a series of well-ordered, equally spaced sawtooth structures,[14-17] they represent the repetitive breaking of "sacrificial bonds" within an intermolecular β-sheet, and the subsequent extension of the "hidden length" of the amino acid chain.[14] This mechanism imparts superior mechanical strength and toughness when compared to systems of similar overall length without such "sacrificial bonds". This is due to the extra energy required to extend the "hidden length" and rupture the "sacrificial bonds" before the main structural backbone is broken.[35] This same mechanism has been measured previously in modular proteins such as titin,[36] and the "glue" adding fracture resistance to bone.[37] It does not have to be associated with adhesion, but instead influences the cohesive strength of the material. Also, it is important to note that in complex materials it is not yet feasible to identify (sub-)components of the material from the mechanical response alone, as the uniqueness of mechanical "signatures" or "fingerprints" is unproven. Such terms may only be valid for synthesized or purified single component systems. The underlying mechanism responsible for the strength of a natural adhesive can thus be deduced from the characteristic mechanical properties of the constituent molecules, rather than from the biochemical identification of specific molecules present.

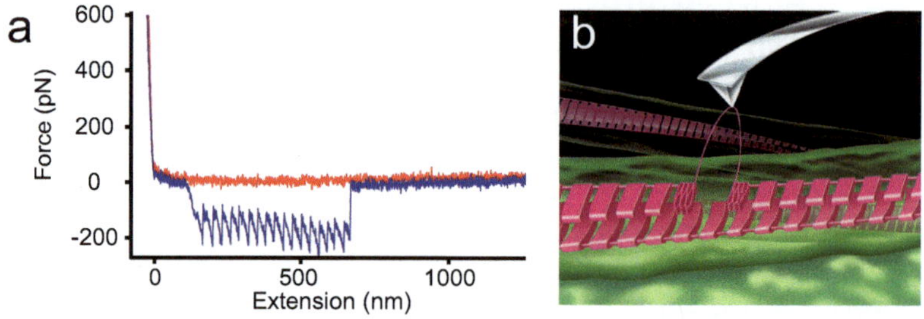

Figure 8.4 Nanoscale mechanical unfolding of amyloid fibril β-sheets in algal adhesive. (a) A representative AFM force–extension curve from holdfasts of the green alga *P. linearis*, showing highly ordered sawtooth structures. (b) Schematic model of the mechanical manipulation of a single intermolecular β-sheet (pink) at the surface of an amyloid fibril (green) by an AFM tip. The schematic shows interacting molecular building blocks of intramolecular β-sheets running perpendicular to the direction of the fibril making up extended intermolecular β-sheets that form the fibril. The "hidden length" is contained within the folds of each individual monomer (detail is shown only below the tip) and the "sacrificial bonds" exist between these folds. Interactions between the monomers must be significantly stronger than these sacrificial bonds and imply cross-linking between each individual protein, as indicated with the solid line running along the back of the manipulated intermolecular β-sheet. With kind permission from Springer Science + Business Media: *Journal of Biological Physics*, Nanoscale mechanical characterisation of amyloid fibrils discovered in a natural adhesive, **32** (2006) 393–401, A. S. Mostaert, M. J. Higgins, T. Fukuma, F. Rindi, S. P. Jarvis, figures 2(a), 5.[14] See also Colour Insert.

The repetitive sawtooth response can be quantitatively analysed using the worm-like chain model for polymers[36,38] that provides, amongst other parameters, a persistence length. In the case of pulling apart the protein monomer constituents of an amyloid fibril, the persistence length approximates the size of an amino acid (this should not be confused with the persistence length of the fibril itself). The contour length corresponds to the "hidden length" exposed on release of the sacrificial bonds within each structural unit, and is directly related to the number of amino acids within this hidden length.

Quantitative variations have been found between functional amyloid from different biological adhesive systems (Fig. 8.5).[17] These are thought to be primarily associated with the differing number of amino acids in the constituent monomers. Where details of the monomer are known, as in the case of curli (see Chapter 7), there is a good correspondence between the quantitative characteristics of the force–extension curve and the anticipated hidden length within the folded monomer.[17] The relative toughness in each case could be calculated from the area under the force–extension curve as this corresponds to the energy required to pull the material apart. The relative toughness of functional amyloid measured *ex vivo* was found to be consistently higher than that measured for amyloid fibrils formed *in vitro*, which may indicate an underlying structural difference between the two.[17] However, in all cases, the molecular level origin of the cohesive strength was associated with the generic amyloid intermolecular β-sheet structure, whereas adhesive residues appeared to lie outside the amyloid core, as deduced by comparing the mechanical responses of amyloid fibrils with and without a specific adhesive function.[17]

An additional material characteristic of amyloid fibrils of benefit to the mechanical properties of the adhesive is the high degree of rotational symmetry provided by the fibril structure itself. Surprisingly, some force traces taken for algal adhesives while approaching the surface have also shown sawtooth structures, indicating unravelling of β-sheets as the tip is pushed into the adhesive matrix (Fig. 8.6).[15] The smaller persistence length on approach, compared to sawtooth structures in the retraction curves, corresponds to the additional stiffness of the matrix material acting in parallel with the fibril. Sawtooth mechanical responses such as these indicate the multi-directional toughness provided by the fibrillar structure. The ability of functional amyloid in natural adhesives to rapidly self-assemble was also apparent in the mechanical data from algae adhesives where sawtooth structures re-formed between successive approach–retraction scans.[15] This rapid reassembly indicates the ability of the material to self-heal.

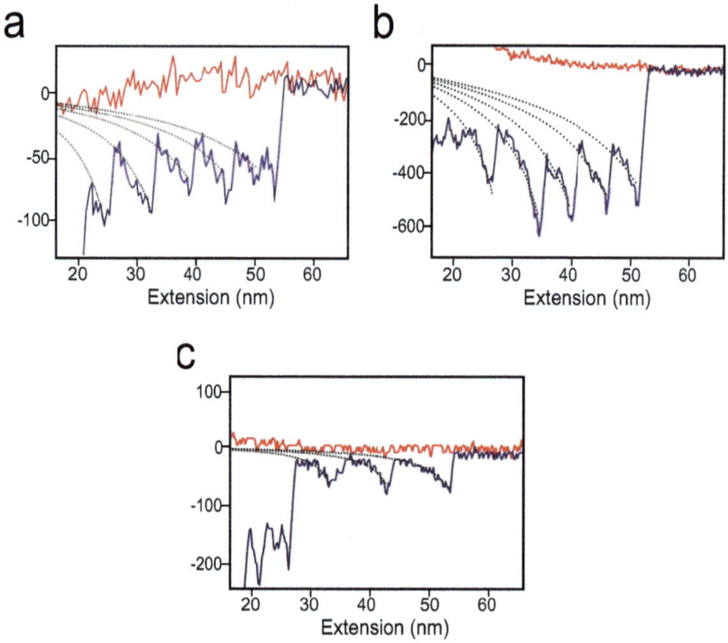

Figure 8.5 Comparison of the sawtooth response from amyloid structures using AFM. (a) Sawtooth response in the force–extension curve from the amyloid fibril in the adhesive of *Entobdella soleae.* (b) Sawtooth response in the force–extension curve from a curli fibril in the adhesive biofilm of *E. coli.* (c) Sawtooth response in a typical force–extension curve from amyloid fibrils grown from α-synuclein. Sawtooth responses are shown fit to the WLC model (dotted lines). Both the approach (red) and retraction (blue) force traces are shown. From A. S. Mostaert, R. Crockett, G. Kearn, I. Cherny, E. Gazit, L. C. Serpell, and S. P. Jarvis, Mechanically functional amyloid fibrils in the adhesive of a marine invertebrate as revealed by Raman spectroscopy and atomic force microscopy, *Archives of Histology and Cytology,* **72**, 199–207 (2009), reprinted by permission of the publisher (International Society of Histology and Cytology).[17] See also Colour Insert.

While the chemical composition of many biological adhesives remains unknown, by using AFM to pull apart individual amyloid fibrils from adhesives, new insights into the underlying mechanical design for adhesive strength and attachment at the molecular level have been provided. These mechanical properties indicate that nature has produced an optimized "glue" that effectively uses mechanically functional amyloid fibrils in biological adhesives and cements for environmental success. This strategy is used by otherwise unrelated organisms including algae,[14–16] a marine invertebrate,[17] and barnacle cement.[21,22] It constitutes the first mechanistic connection identified between the natural adhesives produced by different organisms across different environments.[17]

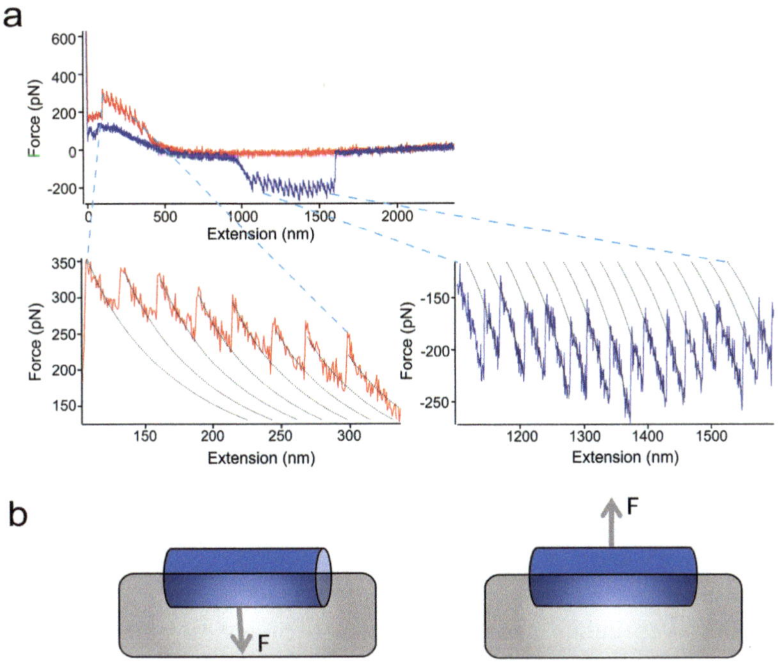

Figure 8.6 AFM force data showing sawtooth structures as the tip approaches the substrate surface through the adhesive and the corresponding schematic. (a) Sawtooth features in the approach and retraction curve. Expansions of the sawtooth regions are shown directly below. A mean persistence length of 0.08 nm and χ^2 of 1.07 was recorded for the approach curve sawtooth peaks, and 0.45 nm for the sawtooth peaks of the retraction curve. An associated schematic is shown in (b) where F indicates the direction of the applied force on the fibril for each expanded curve shown in (a). From A. S. Mostaert and S. P. Jarvis, Beneficial characteristics of mechanically functional amyloid fibrils evolutionarily preserved in natural adhesives, *Nanotechnology*, **18**, 044010 (2007), reprinted by permission of the publisher (IOP Publishing Limited).[15] See also Colour Insert.

8.4 Biomimetic Development of Amyloid-Inspired Adhesives

Amyloid-based biological materials reap the structural and mechanical benefits of the generic amyloid intermolecular β-sheet structure such as high strength, self-healing, and resistance to degradation, while mitigating amyloid toxicity. These features, along with the ability to be formed from many unrelated proteins,[39] indicate possible development of practical materials using a biomimetic approach.

The high performance of amyloid-based adhesives from diverse organisms in different environments provides inspiration for the biomimetic development of adhesives that are markedly different from those currently available commercially. Adhesives that provide strong, rapid, and biocompatible adhesion to wet surfaces for biomedical applications are yet to be achieved. This gap creates the biggest problem in the area of surgical procedures where currently a number of imperfect or compromised products exist. Amyloid-based adhesives have the potential to fill this unmet need.

A task for the future is to produce these valuable biomedical adhesives in an economically feasible way. Production-scale availability of recombinant amyloid proteins, via industrial expression processes, in combination with biological engineering should enable the development of tailored water-impervious, ecologically safe adhesives for which there exist unlimited applications.

However, unlike chemical adhesives, biological adhesives can be remarkably complex, involving a wide range of interactions and components with different functions. This complexity can be daunting for basic and applied research, but the discovery of the relatively simple and robust amyloid in adhesives, that strongly adhere to many different types of surfaces, affords a great deal of commercial opportunity.

8.5 Conclusions

The association of amyloid with bioadhesion highlights the growing trend reflected in this book, showing that amyloid, historically associated with disease, also has non-pathogenic, important functional roles in nature. As the field develops, it is expected to make a substantial contribution to the design of nanoscale structured materials, and in relation to this chapter, there exist endless applications for the commercialization of strong, water-impervious, biocompatible adhesives derived from amyloid. It is clear from this and other chapters in this book that much is yet to be discovered about the formation, functionality and biocompatibility of amyloid in nature. Central to these efforts is the development of protocols to identify, purify and characterize these proteins more simply, and our current focus is on how this may be accomplished from complex biological adhesives and cements.

References

1. Kamino, K. (2008) *Mar. Biotechnol.*, **10**, 111.
2. Silverman, H. G., and Roberto, F. F. (2007) *Mar. Biotechnol.*, **9**, 661.
3. Smith, A. M., and Callow, J. A. (2006) *Biological Adhesives* (Springer, Heidelberg)
4. Waite, J. H. (1983) *Biol. Rev.*, **58**, 209.
5. Waite, J. H., and Tanzer, M. L. (1981) *Science*, **212**, 1038.
6. Wiegemann, M. (2005) *Aquat. Sci.*, **67**, 166.
7. Kamino, K., Inoue, K., Maruyama, T., Takamatsu, N., Harayama, S., and Shizuri, Y. (2000) *J. Biol. Chem.*, **275**, 27360.
8. Jarvis, S. P., and Mostaert, A.S. (2007) *G.I.T. Imaging & Microscopy*, **9**, 25.
9. Chapman , M. R., Robinson, L. S., Pinkner, J. S., Roth, R., Heuser, J., Hammar, M., Normark, S., and Hultgren, S. J. (2002) *Science*, **295**, 851.
10. Barnhart, M. M., and Chapman, M. R. (2006) *Annu. Rev. Microbiol.*, **60**, 131.
11. Otzen, D., and Nielsen, P. H. (2007) *Cell Mol. Life. Sci.*, **65**, 910.
12. Larsen, P., Nielsen, J. L., Dueholm, M. S., Wetzel, R., Otzen, D., and Nielsen, P. H. (2007) *Environ. Microbiol.*, **9**, 3077.
13. Larsen, P., Nielsen, J. L., Otzen, D., and Nielsen, P. H. (2008) *Appl. Environ. Microbiol.*, **74**, 1517.
14. Mostaert, A. S., Higgins, M. J., Fukuma, T., Rindi, F., and Jarvis, S. P. (2006) *J. Biol. Phys.*, **32**, 393.
15. Mostaert, A. S., and Jarvis, S. P. (2007) *Nanotechnology*, **18**, 044010.
16. Mostaert, A. S., Giordani, C., Crockett, R., Karsten, U., Schumann, R., and Jarvis, S. P. (2009) *J. Adhes.*, **85**, 465.
17. Mostaert, A. S., Crockett, R., Kearn, G., Cherny, I., Gazit, E., Serpell, L. C., and Jarvis, S. P. (2009) *Arch. Histol. Cytol.*, **72**, 199.
18. Kearn, G. C., and Evans-Gowing, R. (1998) *Int. J. Parasitol.*, **28**, 1583.

19. Hamwood, T. E., Cribb, B. W., Halliday, J. A., Kearn, G. C., and Whittington, I. D. (2002) *Folia Parasitol*, **49**, 39.
20. Vowles, G. H. and Francis R. J. (2002), in *Theory and Practice of Histological Techniques*, 5th ed. (ed. Bancroft, J. D., and Gamble, M.), Churchill Livingstone, London, pp. 303–324.
21. Sullan, R. M. A., Gunari, N., Tanur, A. E., Yuri, C., Dickinson, G. H., Orihuela, B., Rittschof, D., and Walker, G. C. (2009) *Biofouling*, **25**, 263.
22. Barlow, D. E., Dickinson, G. H., Orihuela, B., Kulp, J. L., Rittschof, D., and Wahl, K. (2010) *Langmuir*, **26**, 6549.
23. Binnig, G., Quate, C. F., and Gerber, C. (1986) *Phys. Rev. Lett.*, **56**, 930.
24. Dufrene, Y. F. (2002) *J. Bacteriol.*, **184**, 5205.
25. Clausen-Schaumann, H., Seitz, M., Krautbauer, R., and Gaub, H. E. (2000) *Curr. Opin. Chem. Biol.*, **4**, 524.
26. Janshoff, A., Neitzert, M., Oberdorfer, Y., and Fuchs, H. (2000) *Angew. Chem. Int. Ed.*, **39**, 3212.
27. Dammer, U., Popescu, O., Wagner, P., Anselmetti, D., Guntherodt, H. J., and Misevic, G. N. (1995) *Science*, **267**, 1173.
28. Smith, B., Schaffer, T., Viani, M., Thompson, J., Frederick, N., Kindt, J., Belcher, A., Stucky, G., Morse, D., and Hansma, P. (1999) *Nature*, **399**, 761.
29. Callow, J. A., Crawford, S. A., Higgins, M. J., Mulvaney, P., and Wetherbee, R. (2000) *Planta*, **211**, 641.
30. Higgins, M. J., Crawford, S. A., Mulvaney, P., and Wetherbee, R. (2002) *Protist*, **153**, 25.
31. Higgins, M. J., Sader, J. E., Mulvaney, P., and Wetherbee, R. (2003) *J. Phycol.*, **39**, 722.
32. Dugdale, T. M., Dagastine, R., Chiovitti, A., Mulvaney, P., and Wetherbee, R. (2005) *Biophys. J.*, **89**, 4252.
33. Dugdale, T. M., Dagastine, R., Chiovitti, A., and Wetherbee, R. (2006) *Biophys. J.*, **90**, 2987.
34. Chiovitti, A., Heraud, P., Dugdale, T. M., Hodson, O. M., Curtain, R. C. A., Dagastine, R. R., Wood, B. K., and Wetherbee, R. (2008) *Soft Matter*, **4**, 811.
35. Smith, B., Schaffer, T., Viani, M., Thompson, J., Frederick, N., Kindt, J., Belcher, A., Stucky, G., Morse, D., and Hansma, P. (1999) *Nature*, **399**, 761.
36. Rief, M., Gautel, M., Oesterhelt, F., Fernandez, J. M., and Gaub, H. E. (1997) *Science*, **276**, 1109.
37. Fantner, G. E., Hassenkam, T., Kindt, J. H., Weaver, J. C., Birkedal, H., Pechenik, L., Cutroni, J. A., Cidade, G. A., Stucky, G. D., Morse, D. E., and Hansma, P. K. (2005) *Nat. Mater.*, **4**, 612.
38. Bustamante, C., Marko, J. F., Siggia, E. D., and Smith, S. (1994) *Science*, **265**, 1599.
39. Chiti, F., and Dobson, C. M. (2006) *Annu. Rev. Biochem.*, **75**, 333.

Chapter 9

Mammalian Functional Amyloid

Douglas M. Fowler,[a] Atanas V. Koulov,[b] Jeffery W. Kelly,[c] and William E. Balch[b]

[a]*Department of Genome Sciences, University of Washington, Seattle, Washington, USA*
[b]*Department of Cell Biology and the Institute for Childhood and Neglected Diseases and*
[c]*Department of Chemistry and the Skaggs Institute of Chemical Biology, the Scripps Research Institute, La Jolla, California, USA*

dfowler@u.washington.edu

Previous chapters in this book have discussed the discovery and characterization of functional amyloid in a wide range of lower organisms. Functional amyloid in humans is of particular interest because of the links between amyloid formation and disease (see Chapter 1). Amyloid formation is a core feature of a large number of human diseases that involve a striking diversity of tissues and physiological processes. Despite decades of effort, no effective treatment exists for amyloid diseases. Understanding the regulatory mechanisms employed to enable the use of amyloid for function, particularly in humans, could illuminate potential therapeutic approaches to amyloid disease. Presently, Functional amyloid is thought to play a role in two processes in humans: haemostasis and melanin synthesis. This chapter will elucidate what is known about the role and regulation of these functional amyloids in the human body.

---

*Note added in press. An additional form of functional human amyloid has recently been reported; Natural Storage of Peptide Hormones, S. K. Maji *et al.*, *Science*, 2009.

The Functional Fold: Amyloid Structures in Nature
Edited by Suzi Jarvis and Anika Mostaert

ISBN 978-981-4267-40-3 (Hardcover), 978-981-4267-62-5 (eBook)
www.panstanford.com

9.1 Amyloid as a Structural and Regulatory Component of Haemostasis

Regulation of blood clot formation and removal is a critical function of the human haemostatic system. Blood clotting is an important part of response to physical injury. Conversely, inappropriate clot formation can lead to thrombus formation and blood vessel blockage. Clot formation and removal is controlled by a series of proteolytic cascades. Clot formation is induced by activation of the Factor XII protease that initiates a proteolytic cascade resulting in the cleavage of fibrinogen to produce fibrin. Fibrin then polymerizes to form a fibrous network that is a major component of blood clots. Clot dissolution and degradation of polymerized fibrin is accomplished by the protease plasmin, which is generated from plasminogen by tissue-type plasminogen activator (tPA). *In vivo*, tPA and plasminogen are recruited to fibrin clots, thereby increasing their effective concentration and leading to plasmin formation and, ultimately, fibrin degradation.

Recent studies suggest that fibrin fibres may comprise, at least in part, an amyloid-like structure. Fibrin fibres have a high β-sheet content[1] and can be stained with Congo red under certain conditions.[2] Raman spectroscopy studies show that fibrin acquires more β-sheet structure upon assembly into fibres, although polarization studies suggest that β-sheets are oriented parallel to the fibrin fibre axis.[3] Additionally, fibrin-derived peptides assemble into amyloid fibres as shown by circular dichroism, X-ray fibre diffraction, and Congo red staining.[2] Despite these suggestive data, the exact structure of fibrin fibres as well as the role that cross-β-sheet motifs play in the assembly and structural integrity of clots remains largely unknown. One possibility is that fibrin fibres contain amyloid-like microdomains for structural or regulatory purposes.

Intriguingly, experiments suggest that such amyloid structures in fibrin fibres could be important in regulating the degradation of clots. Fibrin clots are degraded by the plasmin protease, which is generated from the zymogen plasminogen by tPA. Plasmin generation by tPA is regulated by the recruitment of both plasminogen and tPA to fibrin clots; the increased effective concentration, along with tPA activation, leads to plasmin formation. The amyloid motifs contained within fibrin may be responsible for the stimulation of tPA-mediated plasminogen processing. In fact, amyloid fibres derived from a variety of different sources, including fibrin peptides, can act as a scaffold to activate tPA.[2] These data suggest that cross-β-sheet-based motifs appear to be a defining feature for targeting tPA activity and, therefore, regulating fibrin clot lysis. Thus, amyloid could represent an important element in fibrin coagulation and clot clearance mechanisms.

Interestingly, microbes can apparently exploit the amyloid-based components of the haemostatic system to enhance virulence. Many studies suggest haemostasis, amyloid and micro-organismal pathogenesis are closely related.[4] The notion that amyloid motifs are involved in the formation and disassembly of blood clots illustrates the potential physiological scope and importance of amyloid in humans.

9.2 Amyloid Coordinates Melanin Biosynthesis in Mammals

Amyloid composed of the Pmel17 protein functions in the biosynthesis of melanin in mammals, apparently both facilitating and regulating the chemical reactivity of melanin small molecule quinone precursors.[5] Melanin is present in the skin and eyes, functioning to prevent tissue damage from UV light irradiation and oxidative stress. Melanin synthesis takes place in secretory lysosome organelles called melanosomes, which are generated by highly specialized cells. Melanin is synthesized by the oxidation and polymerization of tyrosine-based indolequinone monomers. Studies of coat colour variation in mice led to the discovery of the Pmel17 protein, which was discovered in connection with the *silver* (*si*) recessive pigmentation defect.[6] *Si*/*si* mice exhibit progressive loss of melanocytes with a concomitant reduction in pigmentation.[6]

Pmel17 is a transmembrane protein that is trafficked to melanosomes where it is responsible for forming fibrous structures[7] (Fig. 9.1). These fibres consist of a fragment of Pmel17, referred to as Mα, derived from the full-length, transmembrane protein by pro-protein convertase (PC) cleavage.[6,8] These Mα fibres appear to be amyloid. *Ex vivo* melanosomes containing Mα fibres have been shown to stain with Congo red and thioflavin S. Furthermore, recombinant Mα (rMα) forms Congo red/thioflavin T-positive, β-sheet rich fibres that exhibit an X-ray diffraction pattern typical of amyloid[5] (Fig. 9.2).

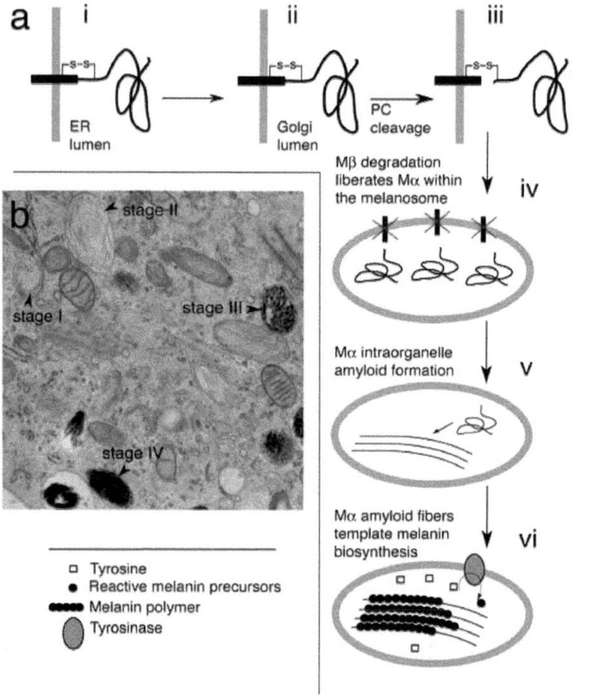

Figure 9.1 Pmel17 trafficking and function in melanin biosynthesis and melanosome biogenesis. (a) Pmel17 is synthesized as a transmembrane protein in the ER (i), it is trafficked first to the Golgi (ii), and finally to the melanosome. Proprotein convertase (PC) cleavage in a post-Golgi compartment creates a lumenal fragment, Mα, and a transmembrane fragment, Mβ. Mα remains disulphide-bonded to Mβ to prohibit aberrant Mα amyloidogenesis (iii). Degradation of Mβ releases Mα from the membrane (iv), enabling Mα to form amyloid fibres within the melanosome organelle (v). Mα amyloid fibres orchestrate the synthesis of melanin from tyrosine-derived reactive indolequinones and protect the melanocyte from melanin-associated toxicity (vi). (b) Electron micrograph depicting the four-stage process of melanosome maturation. Stage I: melanosomes are vesicles of endosomal origin. Stage II: Mα amyloid fibres form within the nascent melanosome. Stage III: melanin appears along Mα fibres. Stage IV: melanin occludes the Mα fibre structure in melanosomes. Part (a) and figure legend reproduced with permission from Elsevier.[5] Electron micrograph was generated by Ilse Hurbain and Graça Raposo.

The function of Mα amyloid within the melanosome appears to be to template and accelerate the chemical synthesis of melanin (Fig. 9.3). Recombinant Mα fibres, as well as Mα derived from biological sources, accelerate the polymerization of small molecule melanogenic precursors such as 5,6-indolequinone into melanin *in vitro*.[5,6,9] Binding 5,6-indolequinone molecules to Mα amyloid likely reduces the entropic requirements for melanin formation by increasing the effective 5,6-indolequinone concentration and by orienting 5,6-indolequinone monomers along the Mα fibre. Amyloid fibres derived from other amyloidogenic polypeptides, such as α-synuclein and Aβ, afford an equal rate enhancement, whereas protein fibres that do not have a cross-β-sheet structure, such as collagen, do not enhance melanin synthesis.[5] These data indicate that the amyloid structure itself is responsible for enhancing the melanin polymerization rate. Thus, Mα amyloid fibres have a clear function within the melanosome, increasing the rate of melanin formation by acting as a multivalent template for 5,6-indolequinone polymerization (Fig. 9.3).

The *si* allele encodes variant of the Pmel17 protein that lacks the cytoplasmic portion of the protein, presumably preventing trafficking from the endoplasmic reticulum to the melanosome. Thus, melanosomes in *si/si* mice do not form Mα fibres. Curiously, they are still able to generate melanin owing to the oxidation and polymerization of melanogenic precursor compounds within the melanosome. However, *si/si* individuals suffer from a progressive loss of melanocytes and cultured *si/si* melanocytes grow more slowly than wild-type melanocytes.[6] These data suggest that Mα amyloid fibres function in protecting melanocytes from the toxic side effects of generating melanin. Melanin precursors including 5,6-indolequinone are extremely cytotoxic; these compounds are potent oxidizers that can react both as electrophiles and nucleophiles.[10] Furthermore, their low charge density makes them likely to diffuse easily across the melanosomal membrane into the cytosol. Such leakage explains the progressive melanocyte destruction observed in the *si/si* animals. Therefore, Mα amyloid appears to bind and sequester reactive melanogenic intermediates, preventing their diffusion out of the melanosome (Fig. 9.4).

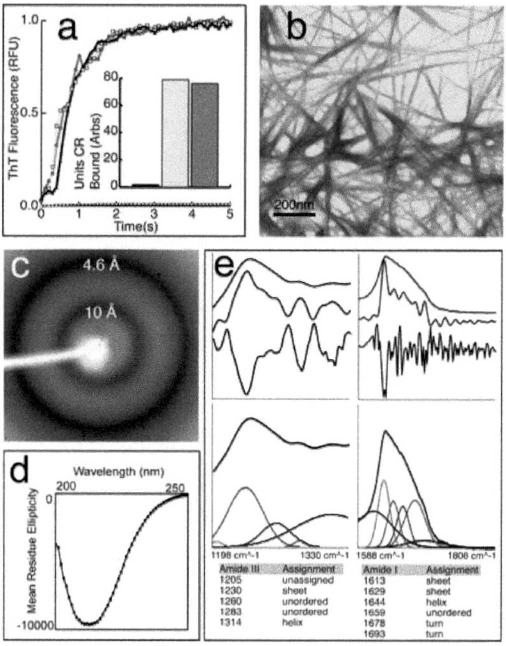

Figure 9.2 Recombinant Mα (rMα) forms amyloid *in vitro*. (a) rMα fibre formation at varying pHs: pH 7.4 (black line, triangles), pH 6.0 (dark grey line, circles), and pH 4.85 (light grey line, squares); control (thioflavin T buffer; black line, white diamonds). The inset bar graph reflects endpoint Congo red binding of equimolar amounts of Mα (dark grey), Aβ 1–40 fibres associated with Alzheimer disease (light grey), and control (Congo red buffer; black). (b) Transmission electron micrograph of typical rMα amyloid fibres with an average diameter of 10 nm. (c) X-ray powder diffraction of lyophilized rMα fibres exhibit a very strong reflection at 4.6 Å and a strong reflection at 10 Å, which is expected of an amyloid cross β-sheet structure. (d) The far-UV CD spectra of soluble Mα aggregates formed at low concentrations to avoid precipitation support a predominantly β-sheet structure. Mα aggregates are approximately 11% α-helix, 32% β-sheet, 23% β-turn, and 33% disordered, based on curve fitting with a basis set of 43 soluble proteins. (e) The attenuated total reflectance FT-IR spectrum of rMα fibres supports a β-sheet-rich structure. Peaks in the amide III (top left, upper curve) and I (top right, upper curve) regions were identified using Fourier self-deconvolution (top left and right, middle curve) and confirmed by second derivative analysis (top left and right, bottom curve). Reproduced from Fowler *et al.*[5]

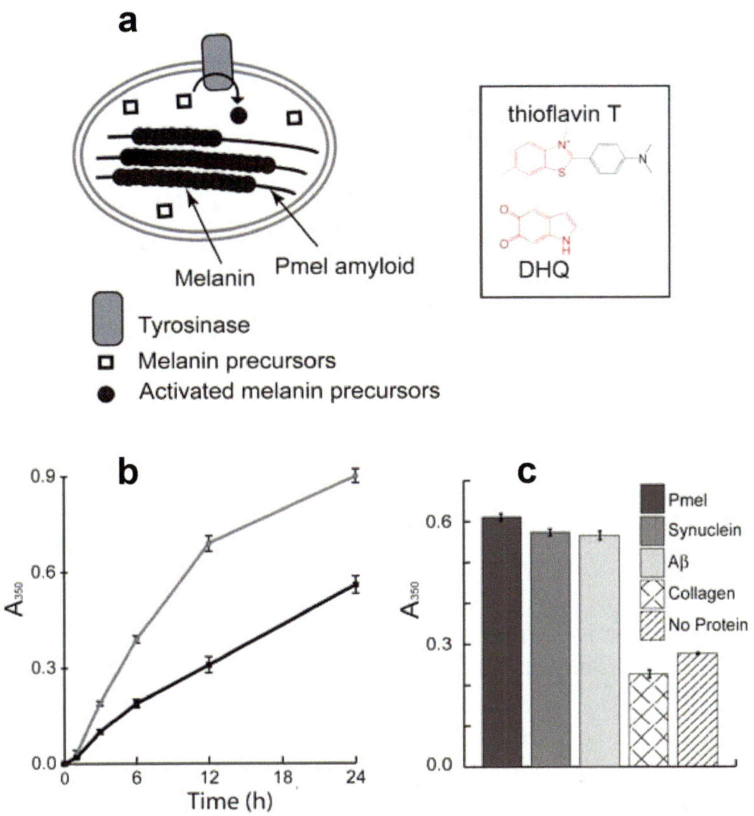

Figure 9.3 Pmel fibres accelerate melanin synthesis. (**a**) In melanosomes, assembly of activated melanin precursors, generated by tyrosinase, occurs along Pmel17 fibres. The boxed portion of (a) illustrates the amyloid-binding dye thioflavin T and the activated melanin precursor 5,6-indolequinone (DHQ), which possess similar core structures. This suggests an explanation for the ability of Pmel17 to concentrate and organize melanin precursors, thereby enabling melanogenesis. (b) A time course of melanin synthesis *in vitro* shows that insoluble rMα amyloid increases the amount of insoluble melanin formed per unit time (grey line) versus a control reaction lacking rMα (black line). (c) Melanin synthesis after 20 h was also evaluated in the presence of insoluble rMα amyloid, α-synuclein amyloid, Aβ amyloid, and collagen IV α-helical fibres. Amyloid fibres, regardless of the component protein, are able to enhance melanin synthesis. Reproduced from Fowler *et al.*[5] See also Colour Insert.

This hypothesis explains observations made in the *si/si* mouse and in *dominant white* (*I*) chicken pigment mutants. The *I* allele is one of the most common chicken pigment variants, and is the result of a polymorphism at the Pmel17 locus.[11] The *I* allele comprises a three amino acid insertion into the transmembrane region of the Pmel17 protein. This insertion likely causes problems with Pmel17 folding and trafficking, meaning *I*/* chickens presumably do not have properly formed Mα fibres. Like *si/si* mouse melanocytes, melanocytes from *I*/* chickens are lost prematurely *in vivo*; when cultured, they die more rapidly than wild-type melanocytes.[12] Additionally, *I*/* melanocytes have significantly diminished superoxide dismutase (SOD) and glutathione (GSH) 1 levels.[13] SOD and GSH restore wild-type-like melanocyte lifespan when added exogenously to the media of cultured *I*/* melanocytes.[13] Melanogenic intermediates including 5,6-indolequinone are powerful oxidizers and their inappropriate presence in the cytosol would be expected to disrupt redox balance within the cell. The oxidative damage and consequent melanocyte death observed in *si/si* mice and *I*/* chickens can be explained by a lack of Mα amyloid fibres, which functions to keep reactive melanogenic precursor compounds such as 5,6-indolequinone within the melanosome.

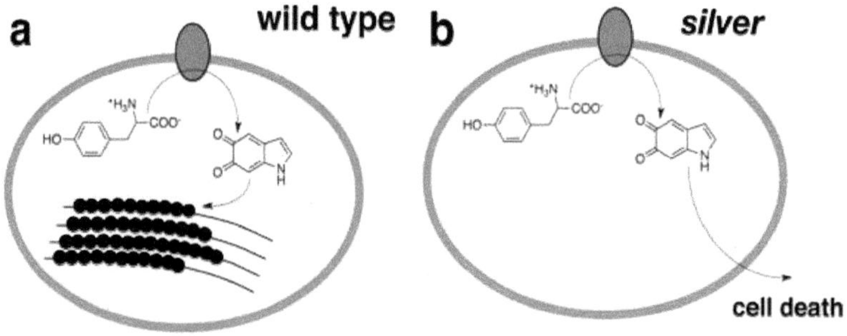

Figure 9.4 Pmel fibres protect the cell from cytotoxic melanogenic intermediates. (a) In a wild-type melanosome, melanogenesis is initiated by the oxidation of tyrosine by tyrosinase. Oxidized melanogenic intermediates polymerize into melanin along Mα amyloid fibres, resulting in mature melanosomes. (b) The *si* Pmel17 mutant is unable to make Mα amyloid fibres. Tyrosinase oxidizes tyrosine; without the Mα fibres to template melanin synthesis, oxidized melanogenic intermediates are free to leak out of the melanosome and cause cell death.

Amyloid formation can be deleterious in many contexts, so it is interesting to note that several features of Pmel17 amyloidogenesis seem designed to avoid toxicity. For example, the Pmel17 protein has many features that prevent inappropriate fibre assembly. The proteolytic cleavage of full-length Pmel17 that enables amyloid formation by the lumenal Mα fragment occurs in post-Golgi compartments that go on to become melanosomes[7] (Fig. 9.2). Because full-length Pmel17 is incapable of forming fibres,[8] cellular compartments upstream of the melanosome, such as the ER and Golgi, are protected from aberrant Mα assembly. After proteolytic cleavage, Mα remains joined by a disulfide bond to the transmembrane/cytoplasmic fragment, Mβ.[7] This could serve as a second physical barrier against improper assembly of Mα. Mα is likely then released from the membrane when Mβ is degraded; this event initiates amyloid formation. Finally, Pmel17 stability and processing appears to be positively influenced by the melanosomal MART-1 protein,[14] suggesting that MART-1 may be a functional regulator of Pmel17 folding and amyloidogenesis.

In addition to an ornate system for preventing premature aggregation, Pmel17 features extremely rapid amyloidogenesis that could be important in avoiding amyloid-related toxicity. Soluble amyloid oligomers appear to be highly toxic.[15] In particular, pore-like amyloid oligomers that compromise membrane integrity have been observed and are hypothesized to be the toxic species responsible for cell and tissue damage in amyloid disease.[16] If formed, Mα amyloid oligomers could rupture the melanosomal membrane and cause leakage of toxic melanogenic precursor compounds into the cytosol. However, oligomeric species are not likely populated to any significant extent during Mα amyloidogenesis because it occurs at least four orders of magnitude faster than pathological Aβ or α-synuclein amyloid formation.[5] Extremely rapid amyloidogenesis, probably the result of an energetically favourable downhill polymerization would minimize the concentration of oligomers and thereby avoid exposing the cell to this key toxic species. By endowing Pmel17 with the ability to form amyloid so rapidly, evolution appears to have circumvented one of the fundamental challenges facing functional amyloid — the generation of toxic oligomeric intermediates. Further study is required to understand the broader principles of regulation of functional amyloidogenesis and the potential interplay between functional and pathological amyloid.

9.3 Conclusion

Functional amyloid in mammals is of particular interest because amyloid deposition is a feature of many poorly understood and presently untreatable human diseases (see Chapter 1). Before more effective therapeutic approaches to amyloid disease can be designed and implemented, a thorough understanding of the mechanisms of amyloid toxicity is necessary. Clues about amyloid toxicity may be gleaned from studying how organisms, particularly higher organisms, regulate and control amyloid formation and degradation. Furthermore, these regulatory mechanisms might be co-opted to treat amyloid diseases.

One interesting question that remains unanswered is how mammalian amyloid is degraded. Fibrin fibres are degraded by the plasmin protease but these fibres likely contain only short amyloid-like structures. Pmel17 forms fibres composed of cross-β-sheets inside melanosomes. The fate of melanosomes in various tissues is the subject of debate; in the skin, melanosomes are passed to keratinocytes. Keratinocytes form the surface of the skin, so in the skin, melanosomes are ultimately shed rather than being degraded. Melanosomes in the eye cannot be shed, and their fate is not well understood. Recent studies have suggested that higher organisms might contain biological pathways for promoting amyloid formation as well as degrading amyloid under pathological conditions.[17] It is possible that such generalized systems exist for regulating the formation and removal of functional amyloid.

In mammals, amyloid-like structures appear to fulfil two very different functional roles — modulating the chemistry of melanin synthesis and helping to regulate haemostasis. The fact that the two known examples of mammalian functional amyloid are so divergent demonstrates the potential range of functions amyloid can fulfil (Fig. 9.5). It also suggests that many more instances of functional amyloid may be found. The discovery of these two examples will hopefully stimulate further efforts to identify and characterize functional amyloid in mammals. Given the diversity and complexity of mammalian cellular physiology, amyloid-like structures may be found with novel and unexpected functions.

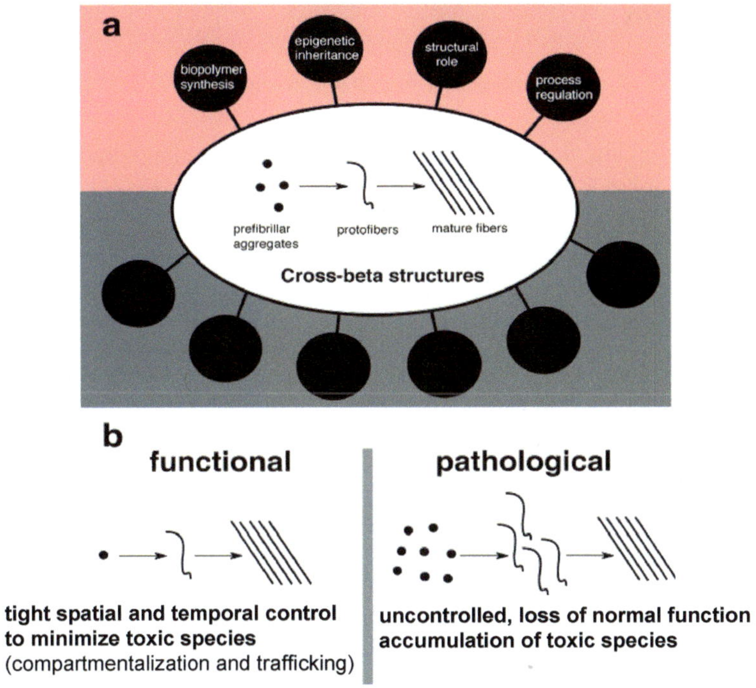

Figure 9.5 Functional amyloid can play a variety of roles and is highly regulated. (a) Amyloid can refer to a diversity of structures that have the cross-β-sheet fold; these range from relatively small pre-fibrillar aggregates to fully formed fibres. Collectively, these cross-β-sheet-rich structures have been shown to perform a wide variety of physiological functions (black circles, top). Given the fact that many polypeptides can form amyloid under the right conditions, it is reasonable to expect that many more instances of functional amyloid are likely to be identified (black circles, bottom). (b) Functional and pathological amyloids share a common cross-β structure and, presumably, a common assembly mechanism. In the case of Pmel17, amyloidogenesis is highly regulated in both space and time to prevent toxicity, whereas pathological amyloidogenesis is uncontrolled and leads to accumulation of toxic species.

Acknowledgement

This work was supported by the National Institutes of Health (GM42336 and GM33301 to WEB and AG18917 to JWK and WEB). AVK is a Fellow of the Cystic Fibrosis Foundation.

References

1. Bramanti, E., Benedetti, E., Sagripanti A., *et al.* (1997) *Biopolymers*, **41**(5), 545.
2. Kranenburg, O., Bouma, B., Kroon-Batenburg, L. M., Reijerkerk, A., We, Y. P., Voest, E. E., and Gebbink, M. F. (2002) *Curr. Biol.*, **12**(21), 1833.
3. Hudry-Clergeon, G., Freyssinet, J. M., Torbet, J., and Marx, J. (1983) *Ann. N.Y. Acad. Sci.*, **408**, 380.
4. Gebbink, M. F., Claessen, D., Bouma, B., Dijkhuizen, L., and Wosten, H. A. (2005) *Nat. Rev.*, **3**(4), 333.
5. Fowler, D. M., Koulov, A. V., Alory-Jost, C., Marks, M. S., Balch, W. E., and Kelly, J. W. (2006) *PLoS Biol.*, **4**, e6.
6. Theos, A. C., Truschel, S. T., Raposo, G., and Marks, M. S. (2005) *Pigment Cell Res.*, **18**(5), 322.
7. Berson, J. F., Harper, D. C., Tenza, D., Raposo, G., and Marks, M. S. (2001) *Mol. Biol. Cell*, **12**(11), 3451.
8. Berson, J. F., Theos, A. C., Harper, D. C., Tenza, D., Raposo, G., and Marks, M. S. (2003) *J. Cell Biol.*, **161**(3), 521.
9. Chakraborty, A. K., Platt, J. T., Kim, K. K., Kwon, B. S., Bettett, D. C., and Pawelek, J. M. (1996) *Eur. J. Biochem.*, **236**(1), 180.
10. Pawelek, J. M., and Lerner, A. B. (1978) *Nature*, **276**(5688), 626.
11. Kerje, S., Sharma, P., Gunnarsson, U., Kim, H., Bagchi, S., Fredriksson, R., Schutz, K., Jensen, P., von Heijne, G., Okimoto, R., and Andersson, L. (2004) *Genetics*, **168**, 1507.
12. Bowers, R. R., Harmon, J., Prescott, S., Asano, J., and Wynne, S. (1992) *Pigment Cell Res.*, **Suppl 2**, 242.
13. Bowers, R. R., Lujan, J., Biboso, A., Kridel, S., and Varkey, C. (1994) *Pigment Cell Res.*, **7**, 409.
14. Hoashi, T., Watabe, H., Muller, J., Yamaguchi, Y., Vieira, W. D., and Hearing, V. J. (2005) *J. Biol. Chem*, **280**(14), 14006.
15. Lashuel, H. A., and Lansbury, P. T. (2006) *Q. Rev. Biophys.*, **39**(2), 167.
16. Lashuel, H. A., Hartley, D., Petre, B. M., Walz, T., and Lansbury, P. T. (2002) *Nature*, **418**(6895), 291.
17. Cohen, E., Bieschke, J., Perciavalle, R.M., Kelly, J. W., and Dillin, A. (2006) *Science*, **313**(5793), 1604.

Chapter 10

Unfolding the Future

Suzanne P. Jarvis and Anika S. Mostaert

UCD Conway Institute of Biomolecular and Biomedical Research,
University College Dublin, Dublin 4, Ireland

suzi.jarvis@ucd.ie; anika.mostaert@ucd.ie

In this chapter we bring together the findings of the previous chapters into an overview of the emerging field of functional amyloid, its hurdles and its potential, as it develops from discovery to understanding and application.

10.1 The Field to Date

The previous chapters have provided an overview of the emerging field of functional amyloid from its inception in 2000 with the identification of physiological amyloid in fungi[1,2] (Chapter 3) and silkworms[3] (Chapter 4). This was shortly followed in 2002 by a similar discovery in bacterial biofilms[4] (Chapter 7). In 2006, specific mechanical functionality was identified and explained relative to the molecular structure of the generic amyloid fibril form found in natural adhesives[5] (Chapter 8) and in the same year

The Functional Fold: Amyloid Structures in Nature
Edited by Suzi Jarvis and Anika Mostaert

ISBN 978-981-4267-40-3 (Hardcover), 978-981-4267-62-5 (eBook)
www.panstanford.com

there was the first report of functional amyloid identified within the human body as a structure to template melanin[6] (Chapter 9). In 2007, the first systematic study of amyloid distribution *in vivo* revealed an unexpected abundance of amyloid in bacterial communities.[7] With a jump in the number of reports of functional amyloid in the past few years,[8–10] the topic is developing into an exciting new multidisciplinary field with broad implications for our understanding of protein function, the evolution of natural materials and the underlying cause of amyloid-based diseases. As the field expands, the many functional attributes of the amyloid fold are expected to be employed beyond their natural manifestation in novel applications. In this final chapter we explore some of these emerging aspects of functional amyloid as well as summarizing the current challenges in the development of the field.

10.2 The Challenge of Identification

Many of the discoveries of functional amyloid to date have occurred via serendipity in scientific research, often in laboratories that had other, initially unrelated projects on pathogenic amyloid, or staff with prior experience in the pathogenic amyloid field. It is thus highly plausible that amyloid occurs far more commonly in nature than is currently apparent, but has remained unidentified. This may primarily be due to continuing problems with definitive identification of amyloid and the difficulties associated with analysing the composition of highly insoluble amyloid fibrils. Currently, there are a broad range of techniques that are used to suggest the presence of amyloid including Thioflavin T fluorescence, Congo red binding, luminescent conjugated polyelectrolyte probes, circular dichroism spectroscopy, Raman spectroscopy, and Fourier transform infrared spectroscopy, fibril observation in electron microscopy or atomic force microscopy images, X-ray fibre diffraction and the use of amyloid specific antibodies. However, no single technique is accepted in isolation as providing a definitive identification of amyloid due to the large number of false positives that can occur and the often highly specific requirements of the sample preparation not to incur a false negative. As a consequence the onus is on the researcher to obtain as many positive indications of amyloid as is possible for a particular sample using a number of the available techniques. This can often be problematic as it is unlikely that access to numerous techniques will be available at a single location, and thus convincing identification of amyloid remains a major hurdle in the development of the field, especially in the case of serendipitous discoveries.

The isolation of fibrils from natural materials is tremendously challenging, unlike the use of fibrils formed *in vitro* from commercially available peptides.[11] In addition, the extreme structural stability of functional amyloid makes it difficult to apply conventional protein science approaches to their purification and characterization. Special techniques such as preparative gel electrophoresis and specific solubilization by organic acids are often required. A disproportionate amount of time can be spent optimizing these steps, severely delaying the subsequent characterization of amyloid. However, this characterization is often a crucial step in proving that the amyloid serves a biological function for the organism, rather than merely the coincidental presence of amyloid in a sample of natural material. Due to the breadth of functions identified to date, it is likely that there are significant structural differences at the molecular level between fibrils that have been harvested from different living organisms. However, the challenges of purification and characterization will need to be overcome before the full extent of these differences can be explored and understood.

10.3 The Latest Discoveries

Several notable discoveries of functional amyloid have taken place while this book has been in the final stages of production. One recent discovery indicates that a broad family of peptide and protein hormones in secretory granules of the endocrine system are stored in the cross-β-sheet rich conformation which defines amyloid.[9] This form serves the biological function of long-term storage at high concentration within mammalian tissue and, as indicated from *in vitro* studies, specifically within the human body. Whilst the finding clearly indicate that functional amyloids in the pituitary and other organs can contribute to normal cell and tissue physiology, the function in this case runs contrary to the widespread notion that amyloid fibrils are very stable and are not thought to release monomers, which would be a prerequisite upon granule secretion. To dispute this notion the authors performed an amyloid release assay for a selection of hormone fibrils showing that all those studied did indeed release monomeric hormone on dilution. Further, the authors established that the released monomers were indeed functional. This highlights the continuing need to reconsider and reinvestigate widespread beliefs in the area of amyloid research. It is also interesting to note that the first paper indicating the distinct molecular organization and possible crystalline structure in secretory granules was published in 1966[12] but a considerable time has lapsed before Maji *et al.*, provided sufficient evidence, using a substantial number of complementary techniques (see Section 10.2), for the structures to be accepted as being in an amyloid form.

In a second "storage" related function *Aplysia* cytoplasmic polyadenylation element binding protein (ApCPEB) was shown to be able to form an amyloidogenic self-sustaining multimer, which appears to contribute to long-term synaptic changes associated with memory storage.[10] This potential function was first proposed by the same authors in 2003 after observing that ApCPEB had properties consistent with it being a prion-like protein when fused to a reporter protein in yeast.[13] Further, in their original paper the authors highlighted that like the fungal prion protein HetS (see Chapter 6), it is the dominant self-perpetuating form that is associated with a gain of function (biochemical activity in the case of ApCPEB). The lack of direct evidence in neurons at that time necessitated substantial additional research to, in the first instance establish that in *Aplysia* sensory neurons, ApCPEB does indeed exist in a monomeric and multimeric state. Further, to establish the functional link the authors showed that the prion-like conversion from a monomeric to a multimeric form was modulated by synaptic stimulation and that ApCPEB is active in the multimeric prion-like state. Finally the authors utilized an antibody that preferentially binds to the multimeric form and diminishes its activity in order to reach the fascinating conclusions that the multimeric amyloid form of Aplysia CPEB is involved in long-term stabilization of activity-dependent change in synaptic efficacy.

10.4 Relevance to the Field of Pathogenic Amyloid

As has been highlighted in many of the previous chapters, amyloid fibrils are notorious for their association with debilitating and often incurable diseases called the Amyloidoses, which include Alzheimer's, Parkinson's, and glaucoma amongst others. These diseases have come under increasingly intense scrutiny in recent years due to the devastating effect they can reap on aging populations. The detrimental effects of amyloid in the human body would appear to take two forms, in one case the volume occupied and mechanical stiffness associated with a high density of amyloid fibrils in tissue, and in the other case the toxic effects on cells of precursor oligomers, which occur on the kinetic pathway from monomer to fibril. With regard to the latter association with disease it has even been proposed that benefit can be derived from the formation of amyloid fibrils because they sequester toxic oligomers of the amyloid protein.[14]

With the recent emergence of functional amyloid in the human body (Chapter 9) it is hoped that there may be enlightening information to be extracted from this and other forms of functional amyloid that can be utilized in studies towards the prevention of amyloid associated with disease. It is apparent that functional amyloid within the human body must be highly regulated to prevent the toxicity usually associated with amyloid formation. In the case of the functional amyloid in skin pigmentation it has been proposed that the rapidity of amyloidogenesis[6] could circumvent the toxicity associated with the intermediate oligomeric stage.[15]

Some functional hormone amyloids were actually found to be moderately toxic to neuronal cells;[9] however, *in vivo* they are encapsulated in lipid membrane which is anticipated to significantly decrease their toxicity. This raises the question as to whether the interaction of the functional hormone amyloids with lipid membrane occurs specifically with an as of yet unidentified granule-recruiting membrane protein or whether the attraction occurs spontaneously. In either case it seems plausible that the mediation of this lipid in any subsequent cell–amyloid interaction could serve the purpose of decreasing toxicity. The interactions of amyloid at the membrane–fluid interface could thus play a central role in determining amyloid function or provide a pathogenic mechanism. Membrane composition varies greatly between the many organisms which utilize functional amyloid and the cholesterol rich regions of the human body most closely connected with amyloid-based diseases such as the brain and associated neurodegenerative diseases, such as Alzheimer's and Parkinson's.

It is clear that when considering the functional or pathogenic nature of amyloid, the local environment will be a critical determining factor. In other words, primarily functional amyloid can be altered under certain conditions to become disease-associated amyloid. As has been highlighted by Maji *et al.*,[9] several endocrine hormones that may be stored in an amyloid-like state are present as amyloids in amyloid diseases such as amylin, which is associated with type II diabetes. Cross-seeding of amyloids between different protein species have also been reported.[16] These issues are likely to be considered in significantly more depth as researchers attempt to utilize amyloid as a functional biomaterial within the human body.

10.5 Applications of Functional Amyloid

Prior to the recent emergence of widespread functional amyloid, amyloid fibrils were already being considered as a useful material for applications in nanotechnology and bioengineering. This is due to their generic characteristics of chemical, mechanical, and thermal stability as well as their ability to readily self-assemble. Demonstrated potential uses have included implating metal nanowires[17,18] and tissue engineering scaffolds, to promote cell adhesion and growth.[19]

The discovery of a broad range of functional amyloid materials further extends the possible range of applications via the mimicking of these structures to produce a new generation of functional biomaterials. On the basis of nature's own application of functional amyloid, potential biomimetic uses include lubricants (Chapter 2), adhesives (Chapter 8), silks (Chapters 4 and 5), and coatings (Chapter 3). In a number of these applications the mechanical properties of the fibrils (or in some cases fibre composite materials) are of primary functional importance. Nanomechanical measurements have helped shed light on the beneficial mechanical characteristics of functional amyloid fibrils[20] (Chapter 8). It has also been shown that amyloid fibrils assembled *in vitro* from short peptide segments have measured rigidities close to the maximal values expected for defect-free structures of near crystalline order.[21] However, in the design of lubricants and adhesives, the mechanical characteristics of crystalline materials are not those which the materials scientist would select and it remains to be established whether the defects often observed in *ex vivo* fibrils, such as wild-type curli, play a positive role in the yield characteristics of the material.

Other applications, which have not yet been found in nature, can also be envisaged by directing the functional properties into new areas. On such example is the suggested use of amyloid as a depot for the formulation of long-acting drugs.[8] This is based on the rationale that an amyloid forming drug takes the structural form of a stable reservoir, containing only the peptide of interest, whereby drug release is controlled from the fibril termini and the drug is active upon release. The viability of this approach was explored for a family of short- and long-acting analogues of gonadotropin-releasing hormone (GnRH).[8] *In vitro* analysis showed that all of the long-acting analogues formed fibrils that slowly released the amyloid monomers. Further, *in vivo* subcutaneous administration of a fibrillar sample was found to result in a longer duration of action than its corresponding soluble counterpart. It is thought that monomer release *in vivo* may be altered by binding components from the host, such as glycosaminoglycans, that have the effect of stabilizing amyloid fibrils, protecting them from degradation and possibly reducing the potential immune response by covering the fibrils. This would be of considerable practical benefit in applications. The researchers also explored the potential pitfall in attempting to utilize such a drug delivery system of cross-seeding, using one of the long-acting GnRH analogues, and found no evidence of cross-seeding either for wild-type GnRH or α-synuclein.

For applications within the human body, the issue of potential toxicity of these functional fibrils will need to be addressed, as indeed is the case for all new nanomaterials.[22] Here it is important to remember that it is not the material per se that is either toxic or non-toxic but the material in concert with its local environment and its kinetic evolution which need to be considered, understood, and controlled.

10.6 Conclusions

The rapid growth in the discovery of functional amyloid in recent years should help to establish where and why this generic structure is most likely to be found in a functional context. The diversity of functions and organisms utilizing amyloid structures should help considerably in pinpointing the driving forces behind the regulated and unregulated formation of these structures. This knowledge will be utilized for producing novel functional biomaterials and tailored nanostructured amyloid-inorganic composites as well as adding to the research efforts to find preventions and cures for the range of diseases associated with amyloid fibril formation.

References

1. de Vocht, M. L., Reviakine, I., Wösten, H. A. B., Brisson, A., Wessels, J. G. H., and Robillard, G. T. (2000) *J. Biol. Chem.*, **275**, 28428.
2. Wösten, H. A. B., and de Vocht, M. L. (2000) *Biochim. Biophys. Acta*, **169**, 79.
3. Iconomidou, V. A., Vriend, G., and Hamodrakas, S. J. (2000) *FEBS Lett.*, **479**, 141.
4. Chapman, M. R., Robinson, L. S., Pinkner, J. S., Roth, R., Heuser, J., Hammer, M., Normark, S., and Hultgren, S. J. (2002) *Science*, **295**, 851.
5. Mostaert, A. S., Higgins, M. J., Fukuma, T., Rindi, F., and Jarvis, S. P. (2006) *J. Biol. Phys.*, **32**, 393.
6. Fowler, D. M., Koulov, A. V., Alory-Jost, C., Marks, M. S., Balch, W. E., and Kelly, J. W. (2006) *PLoS Biol.*, **4**, e6.
7. Larsen, P., Nielsen, J. L., Dueholm, M. S., Wetzel, R., Otzen, D., and Nielsen, P. H. (2007) *Environ. Microbiol.*, **9**, 3077.
8. Wang, L., Maji, S. K., Sawaya, M. R., Eisenberg, D., and Riek, R. (2008) *PLoS Biol.*, **6**, e195.
9. Maji, S. K., Perrin, M. H., Sawaya, M. R., Jessberger, S., Vadodaria, K., Rissman, R. A., Singru, P. S., Nilsson, K. P. R., Simon, R., Schubert, D., Eisenberg, D., Rivier, J., Sawchenko, P., Vale, W., and Riek, R. (2009) *Science*, **325**, 328.
10. Si, K., Choi, Y.-B., White-Grindley, E., Majumdar, A., and Kandel, E. R. (2010) *Cell*, **140**, 421.
11. Otzen, D., and Nielsen, P.H. (2008) *Cell Mol. Life Sci.* **65**, 910.
12. Miller, F., de Harven, E., and Palade, G. E. (1966) *J. Cell Biol.*, **31**, 349.
13. Si, K., Lindquist, S., and Kandel, E. R. (2003) *Cell*, **115**, 879.
14. Hardy, J., and Selkoe, D. J. (2002) *Science*, **297**, 353.
15. Caughey, B., and Lansbury, P. T. (2003) *Annu. Rev. Neurosci.*, **26**, 267.
16. Han, H. Y., Weinreb, P. H., and Lansbury, P. T. (1995) *Chem. Biol.*, **2**, 163.
17. Scheibel, T., Parthasarathy, R., Sawicki, G., Lin, X.-M., Jaeger, H., and Lindquist, S. L. (2003) *Proc. Natl. Acad. Sci. USA*, **100**, 4527.

18. Reches, M., and Gazit, E. (2003) *Science*, **300**, 625.
19. Holmes, T.C., de Lacalle, S., Su, X., Liu, G. S., Rich, A., and Zhang, S. G. (2000) *Proc. Natl. Acad. Sci. USA*, **97**, 6728.
20. Mostaert, A. S., and Jarvis, S. P. (2007) *Nanotechnology*, **18**, 044010.
21. Knowles, T. P., Fitzpatrick, A. W., Meehan, S., Mott, H. R., Vendruscolo, M., Dobson, C. M., and Welland, M. E. (2007) *Science*, **318**, 1900.
22. Nel, A., Xia, T., Mädler, L., and Li, N. (2006) *Science*, **311**, 622.

Index

Colour Insert

Chapter 1

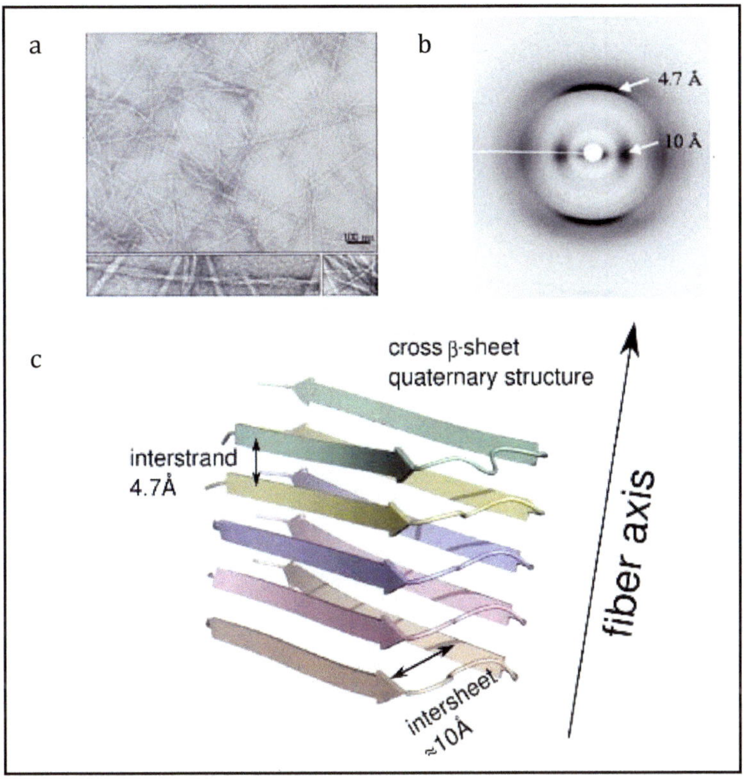

Figure 1.1

Chapter 2

Figure 2.1

Figure 2.3

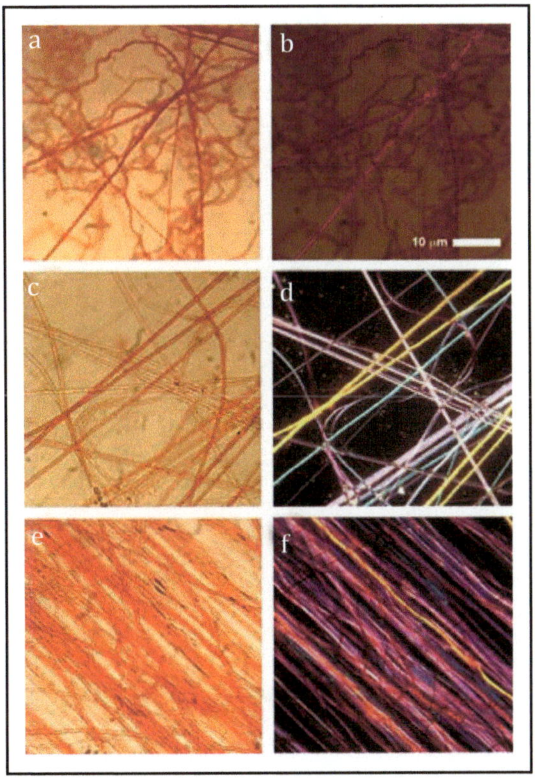

Figure 2.4

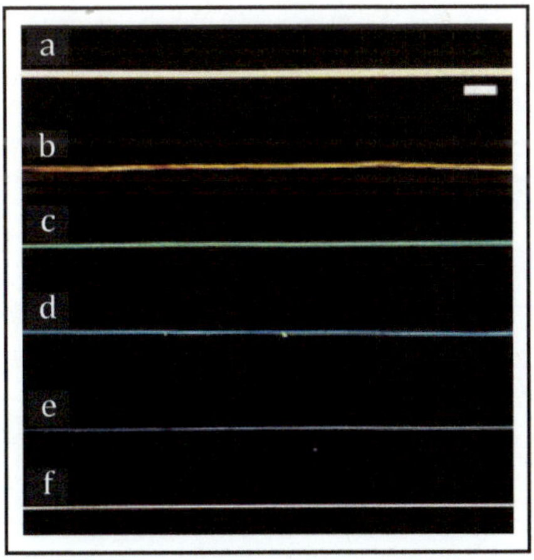

Figure 2.7

Chapter 3

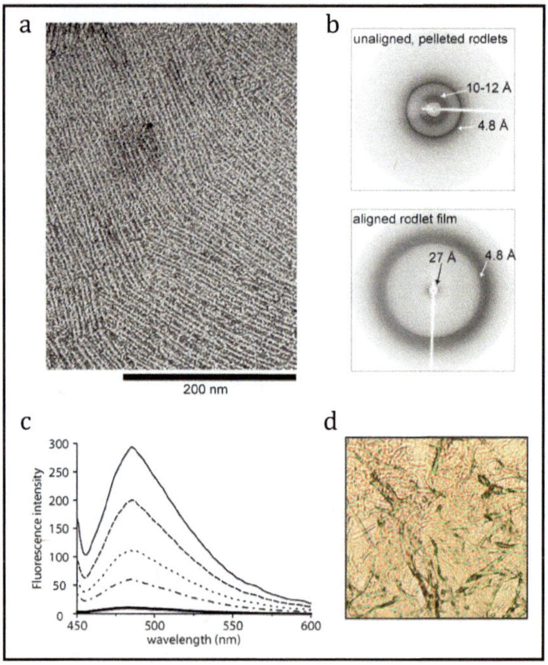

Figure 3.4

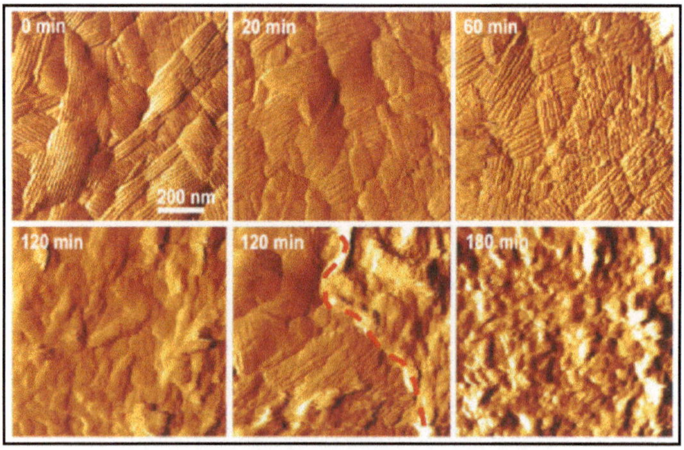

Figure 3.5

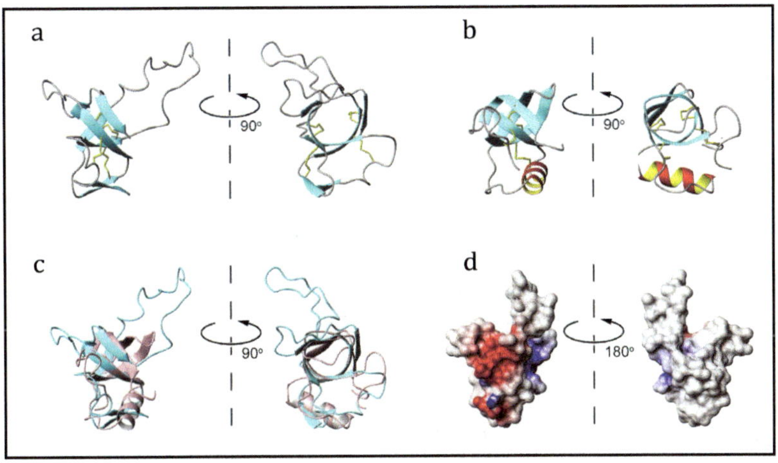

Figure 3.6

Chapter 4

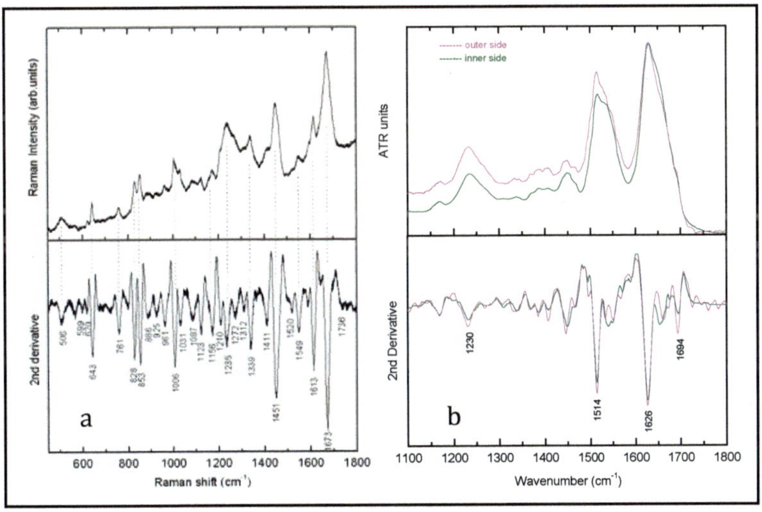

Figure 4.5

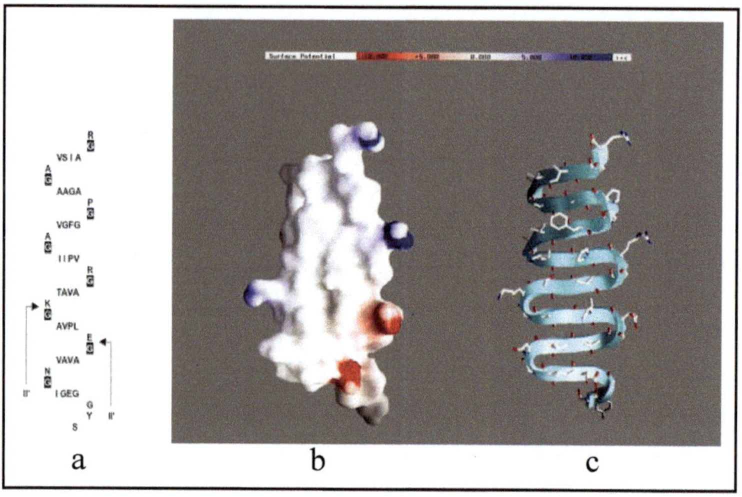

Figure 4.11

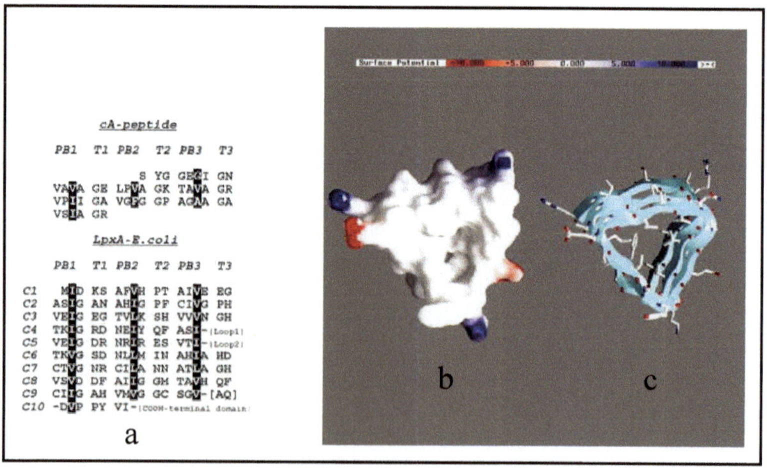

Figure 4.12

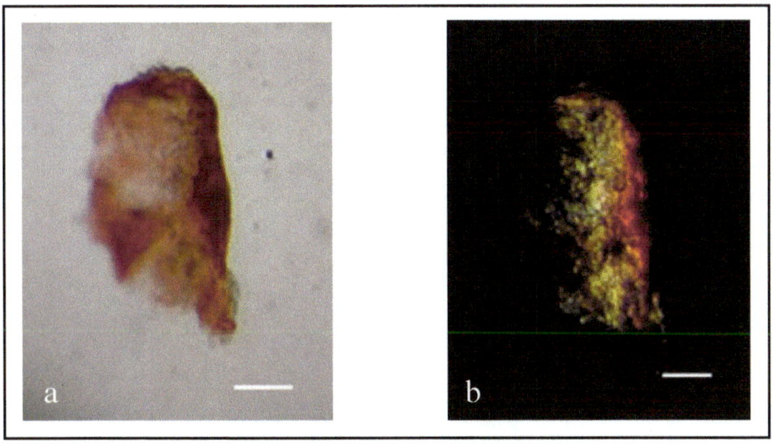

Figure 4.13

Chapter 5

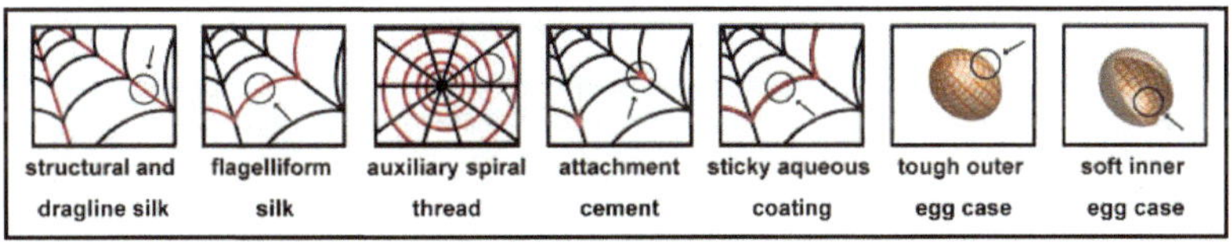

Figure 5.1

Chapter 6

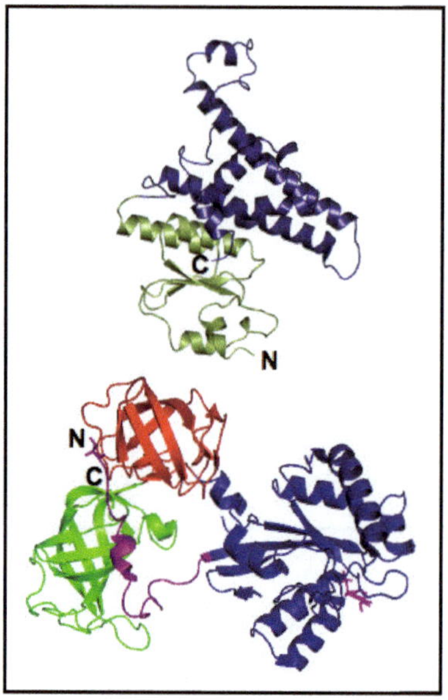

Figure 6.1

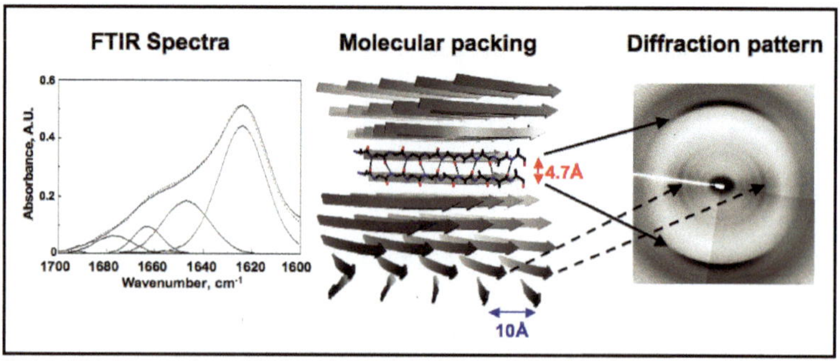

Figure 6.3

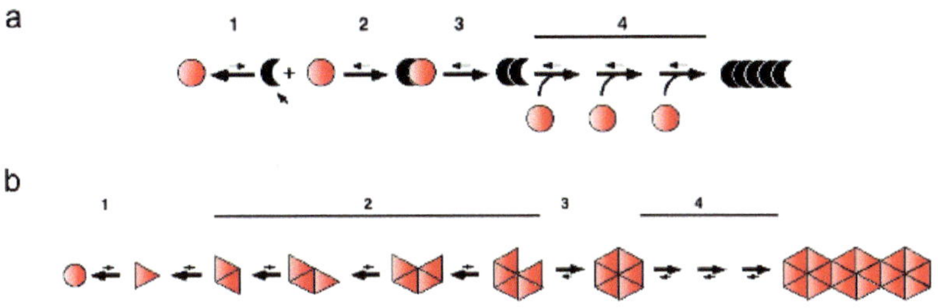

Figure 6.4

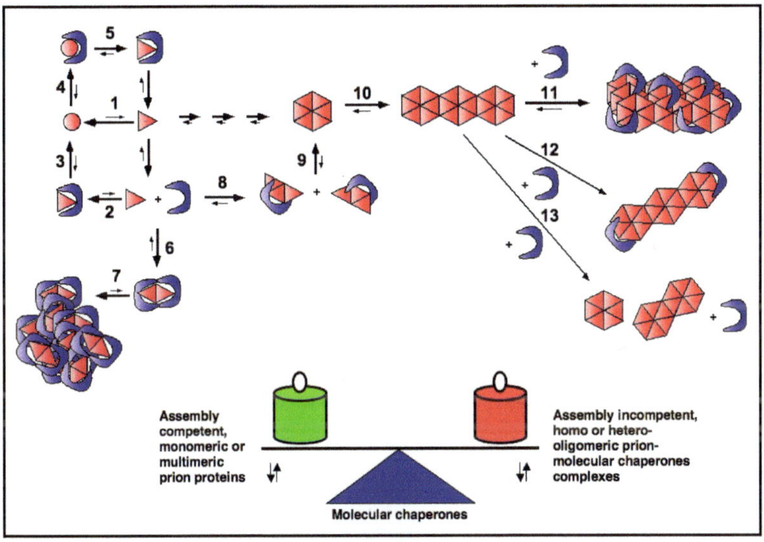

Figure 6.5

Chapter 7

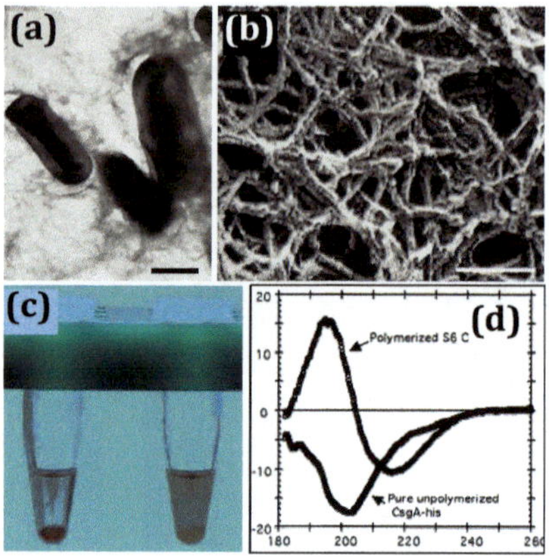

Figure 7.2

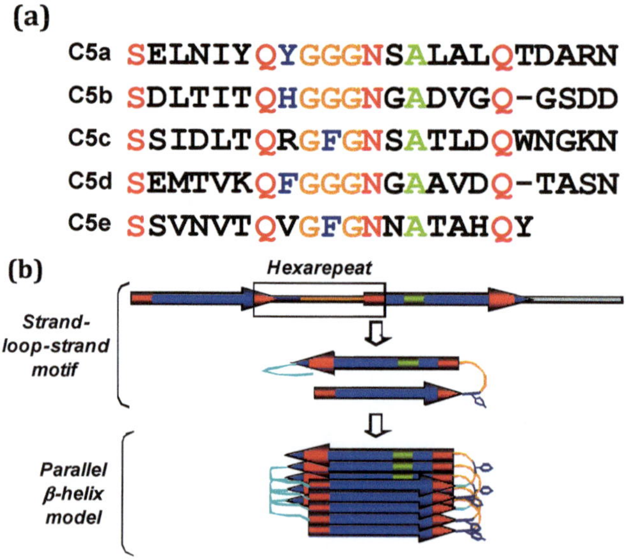

Figure 7.3

Chapter 8

Figure 8.1

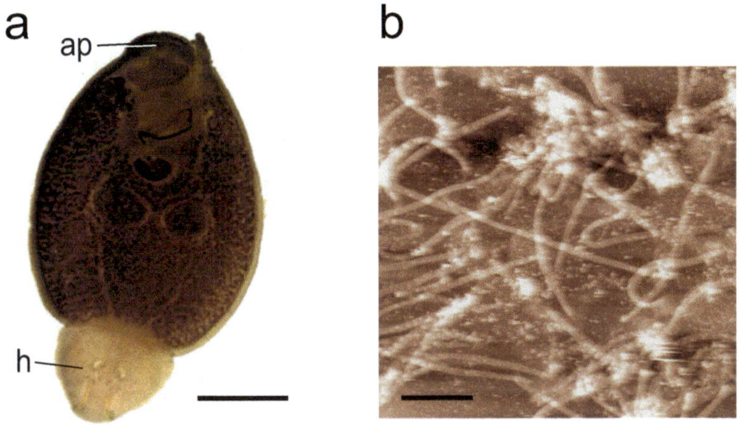

Figure 8.2

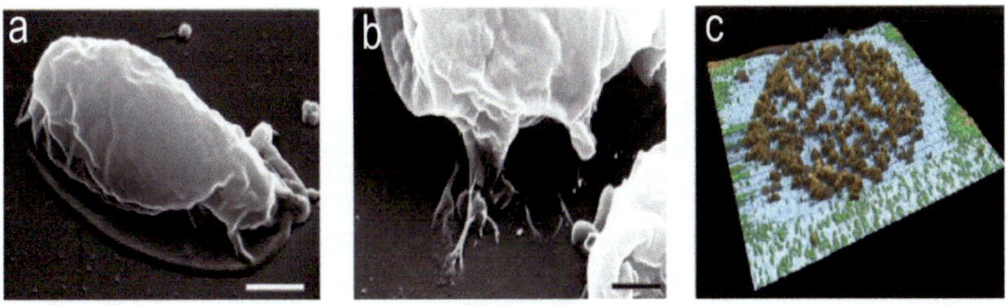

Figure 8.3

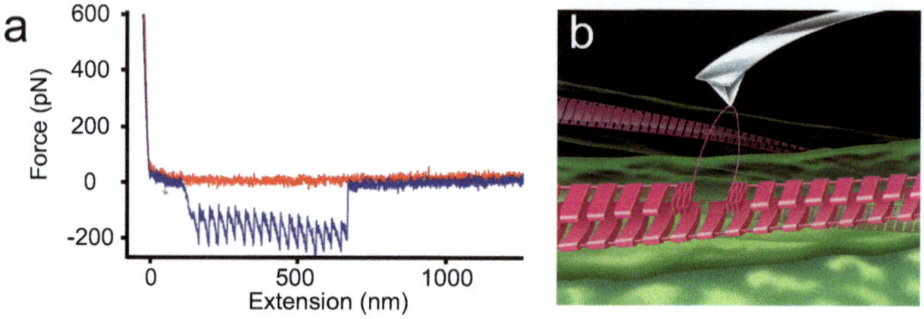

Figure 8.4

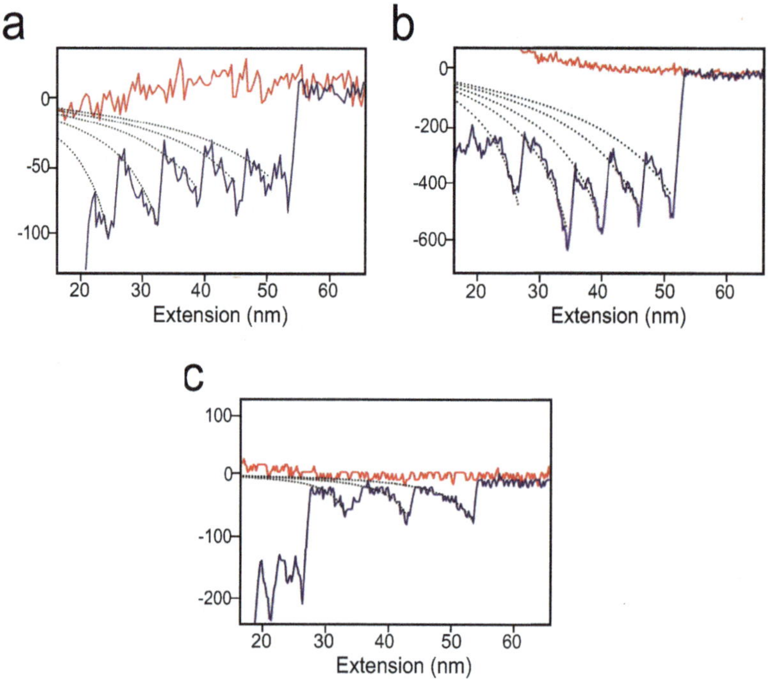

Figure 8.5

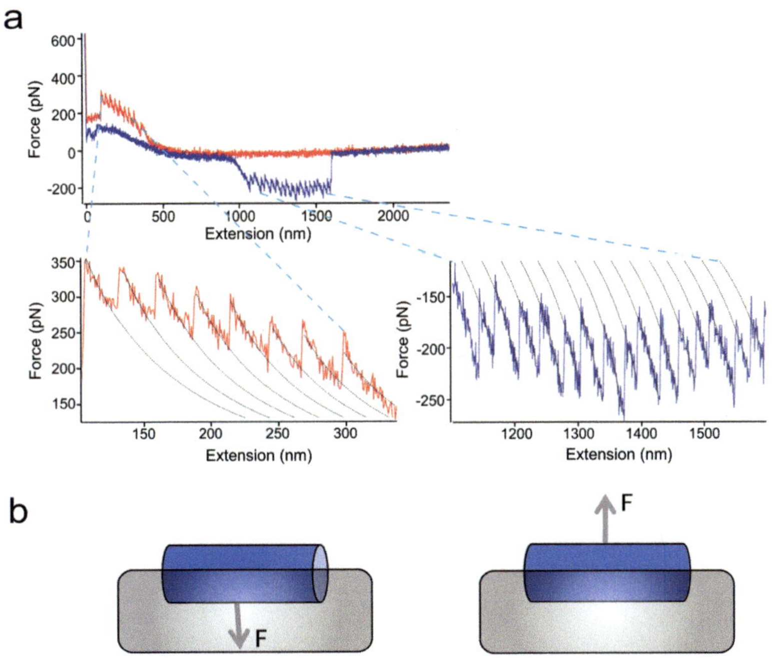
a
Force (pN)
600
400
200
0
-200
0
500
1000
1500
2000
Extension (nm)
350
300
250
200
150
150
200
250
300
Extension (nm)
-150
-200
-250
1200
1300
1400
1500
Extension (nm)
b
F
F

Figure 8.6

Chapter 9

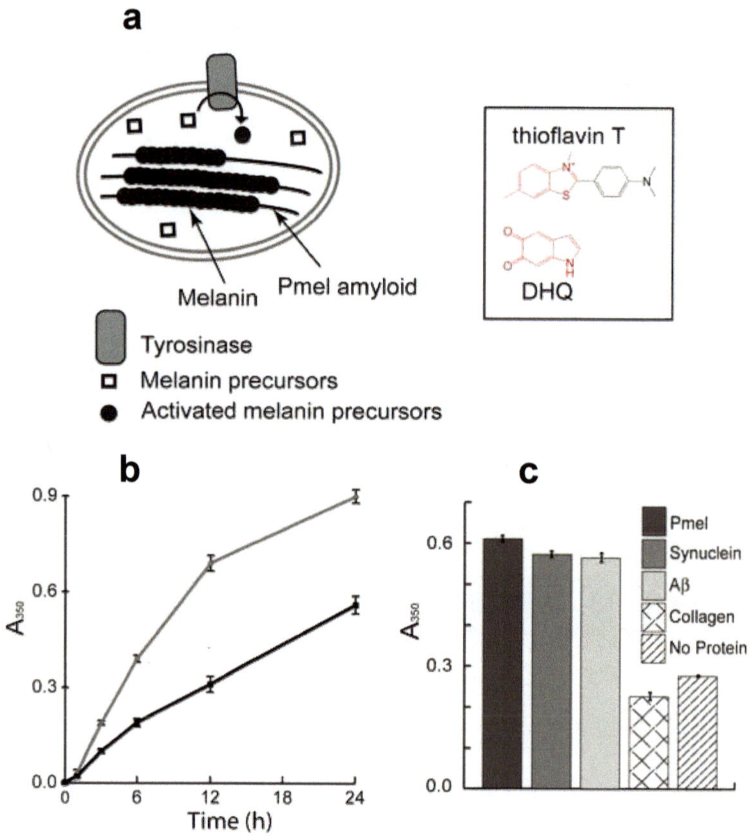

Figure 9.3